S0-BZD-678

Fractals

Fractals

Non-integral dimensions and applications

Under the direction of

G. Cherbit
University of Paris VII

Foreword by

J.-P. Kahane

Translated by

F. Jellett

Financial assistance for the translation was given by the French Ministry of Culture

JOHN WILEY & SONS
Chichester • New York • Brisbane • Toronto • Singapore

Originally published as *Fractals: Dimensions non-entières et applications*

Other Wiley Editorial Offices

John Wiley & Sons, Inc., 605 Third Avenue,
New York, NY 10158-0012, USA

Jacaranda Wiley Ltd, G.P.O. Box 859, Brisbane,
Queensland 4001, Australia

John Wiley & Sons (Canada) Ltd, 22 Worcester Road,
Rexdale, Ontario M9W 1L1, Canada

John Wiley & Sons (SEA) Pte Ltd, 37 Jalan Pemimpin 05-04,
Block B, Union Industrial Building, Singapore 2057

***Library of Congress Cataloging-in-Publication Data*:**

Fractals : non-integral dimensions and applications / under the direction of G. Cherbit ; foreword by J. P. Kahane.
p. cm.
Translation of: Fractals.
Includes bibliographical references and index.
ISBN 0 471 92798 8
1. Fractals. I. Cherbit, G. (Guy)
QA614.86.F74 1990 90-12468
514'.74—dc20 CIP

British Library Cataloguing in Publication Data:

Fractals : non-integral dimensions and applications.
1. Mathematics. Fractals sets. Geometric aspects
I. Cherbit, G.
515.73

ISBN 0 471 92798 8

Typeset by Thomson Press (India) Ltd, New Delhi
Printed in Great Britain by Biddles Ltd. Guildford

We dedicate this work
to the first Magdalenian man
who, in the habit of *counting* his hunting trophies
by means of handprints
on the wall of his grotto
one day had to *share* a catch...

Dr. Paul Mongré

Félix Hausdorff

Contents

Foreword

Jean-Pierre Kahane
University of Paris-sud Orsay

Science, indeed humanity itself, is seriously threatened by the fragmentation of knowledge into myriad disparate specialisms. It is a major problem to fill the gap between advances in all sciences and society's failure to assimilate this progress and its implications. Fortunately, there are signs that matters can be put right, that communication can be established between researchers in different fields, that a vibrant scientific culture can be created from the blending of a variety of disciplines and that science can advance arousing fresh curiosity and stimulating new interests by means of new collaborations and a fresh look at the past.

The present book bears witness to this, and will be a valuable tool and reference work for all those interested in fractal forms and their properties. As varied as it is, it cannot convey the atmosphere generated by the 'Hausdorff seminars': the mathematicians at the blackboard, the physicists with their slides and films, the barrage of questions, the small packed room with nowhere to put one's coat, the apparent disorder and the underlying order, and the common goal: to understand and to be understood.

The organizers were physicists and it was they who chose the name of Hausdorff for their seminar. This was their way of showing the esteem in which they held mathematics. I would like to take the opportunity afforded me by this preface to say something about the mathematics in question—measure theory and Hausdorff dimension—and also about the little known but fascinating person of Felix Hausdorff.

In 1918, Félix Hausdorff was fifty. He had just published a scientific classic *Grundzüger der Mengenlehre* (*Set Theory*); he had made important contributions to topology, algebra and analysis with more to follow in the course of his long career. His theory of measure and dimension comprises a 22-page article, written in March 1918 and immediately published by Mathematische Annalen. He begins by paying tribute to Caratheodory who, in 1914, gave a new treatment of Lebesgue measure, in which the axioms are prominent. 'Hierzu,' he says, 'geben wir im folgendem einen kleinen Beitrag.'* This 'little contribution' is his entire theory of measure and of 'fractional' dimension, presented in a clearer and more general form than most of the existing treatments. He starts with a family $\mathscr{U}$ of

*'In addition', he says, 'we present, in what follows, our own little contribution'

bounded subsets of $\mathbf{R}^q$ and another subset A of $\mathbf{R}^q$. He supposes that A can be covered by at most countably many sets U from the family $\mathscr{U}$ of arbitrarily small diameter $d(U)$. With each U he associates a weight $l(U)$, and with each covering of A by sets U_n of diameters less than ρ he associates the (possibly infinite) sum $\sum l(U_n)$. The infimum of these sums depends on A and ρ; let it be denoted by $L_\rho(A)$. As ρ decreases, fewer coverings are allowed so that the infimum $L_\rho(A)$ can only increase. Its limit as ρ tends to 0, $L(A)$, is an 'outer measure' in the sense of Caratheodory—a notion introduced by Hausdorff at the beginning of his article—satisfying certain very natural conditions set out by Hausdorff. This is indeed a natural and general way of measuring a set A using a collection of subsets $\mathscr{U}$. Hausdorff gives a variety of examples (not at all well-known nowadays) and then addresses himself to the principle application, namely when $\mathscr{U}$ includes all bounded sets and $l(U)$ is a function of the diameter $\rho(U)$, $= \lambda(\rho(U))$. When $\lambda(x) = x^a$, $L(A)$ is what is known today as an a-dimensional measure. This is the point at which Hausdorff defines his 'fractional dimension' (our Hausdorff dimension) and concerns himself with the scale defined by the functions $\lambda(x)$. He shows, for example, constructing suitable Cantor-type sets, that to each concave function $\lambda(x)$ there corresponds an A such that $0 < L(A) < \infty$.

This article is a gem, but it has an odd history: few people have read it, yet it has brought its author more fame, today, than all the rest of his works put together. In its day (1920) it was respected—and followed important work in potential theory, Fourier analysis and, later on, probability theory. However, its basic elementary geometric character was not entirely in keeping with the spirit of the times. Following Lebesgue, measure theory has permeated, via the Lebesgue integral, the whole of classical analysis and contributed to the foundation of functional analysis and then to modern probability theory. Following Cantor and Dedekind, Brouwer and Hausdorff himself, dimension became a primarily topological notion. The geometrical aspect of measure theory—Lebesgue's starting point—gave place to abstract measures, well suited for probability, while the metric aspect of dimension, despite several applications, remained in obscurity until the 1950s.

Today the situation is reversed, as witnessed by the success of fractal geometry. Mathematicians are once again interested in geometrical aspects of measure theory and in metric dimensions. Moreover, the name of Hausdorff already associated by mathematicians with set theory, separability in topological spaces and the moment problem, as well as a great many theorems, is now primarily associated with his 1918 article 'Dimension und äusseres Mass' (Dimension and outer measure').

Nevertheless, Félix Hausdorff is still a relatively little-known figure. However, he personifies several aspects of German life and culture, and of its struggle with the barbarous non-culture of Nazism. Born in Breslau into a rich Jewish family, he developed a taste for music, but under parental pressure, however, he studied science. Then, on completing his studies, he led a double life. As a man of science, starting with astronomy and optics (his degree thesis, in 1895, dealt with the absorption of light in the atmosphere), he duly progressed to pure mathematics.

In artistic and literary circles, however, he was Doctor Paul Mongré, author of poems, philosophical works, various essays and even a successful stage play, a man of wit and culture and known as such. The 1914 war killed off Paul Mongré, leaving just a great mathematician, active and respected.

In due course Nazism arrived, followed by the Second World War. On 26 January 1942, Félix Hausdorff, together with his wife, committed suicide to avoid internment. After the War, over a thousand pages of manuscript written during the dark years were discovered. These were edited in 1969—a massive task of reviewing and collating. The editor was faithful to the original manuscripts, as clear and polished as the thoughts of Félix Hausdorff.

Many thanks to the 'Hausdorff seminarists' for giving me the opportunity to evoke his memory.

J.-P. Kahane
Orsay, 13 November 1986

Chapter 1 Introduction

Guy Cherbit
Membrane Biophysics Research Group, University of Paris VII

Who among us has not been puzzled and then fascinated by Benoît Mandelbrot's astonishing work *Objets Fractals*, published in 1975? With enthusiasm and a fertile imagination, the author revealed all that could be extracted from mathematics, until then familiar only to specialists.

Subsequently, the past decade has seen an almost exponential growth in the number of publications in this field. Applications have evolved in almost all areas and the mathematicians have been set fresh problems.

Although he coined the word 'fractal', B. Mandelbrot refuses to provide either a definition or directions for use. This has the advantage of not trapping the notion in a narrow context and has led to a very rich scientific harvest. However, as he himself writes, this has given rise to a certain amount of confusion among the physicists and suspicion among mathematicians.

Nevertheless, mathematical work on non-integral dimensions and on curves without tangents has proliferated, although one only very occasionally comes across the term *fractal* in such works.

Convinced for many years that the notion of non-integral dimension should be fruitful for theoretical physics and not finding an adequate tool in the Hausdorff dimension, I opted for an experimental approach and developed instrumentation to explore the widest possible range in both time and space (time resolution of the laser spectroscopy of biomolecules).

On the publication of B. Mandelbrot's work, a great many researchers understood the great interest of fractals. However, having read and re-read the work (and its sequels) many times and carefully studied other publications on fractals, many of them tried, by means of the theory, to set up an immediately usable tool.

In this spirit, we organized, on a monthly basis, 'Hausdorff seminars on the notion of non-integral dimension and its applications', whose goal is better to tackle the problems confronting practical scientists.

The present book is not a detailed account of these seminars, but captures, with illustrative examples, the main themes of each session.

I thank all who have taken part in this work, as well as P. Brandouy who acquitted himself with efficiency and enthusiasm in the onerous job of seminar

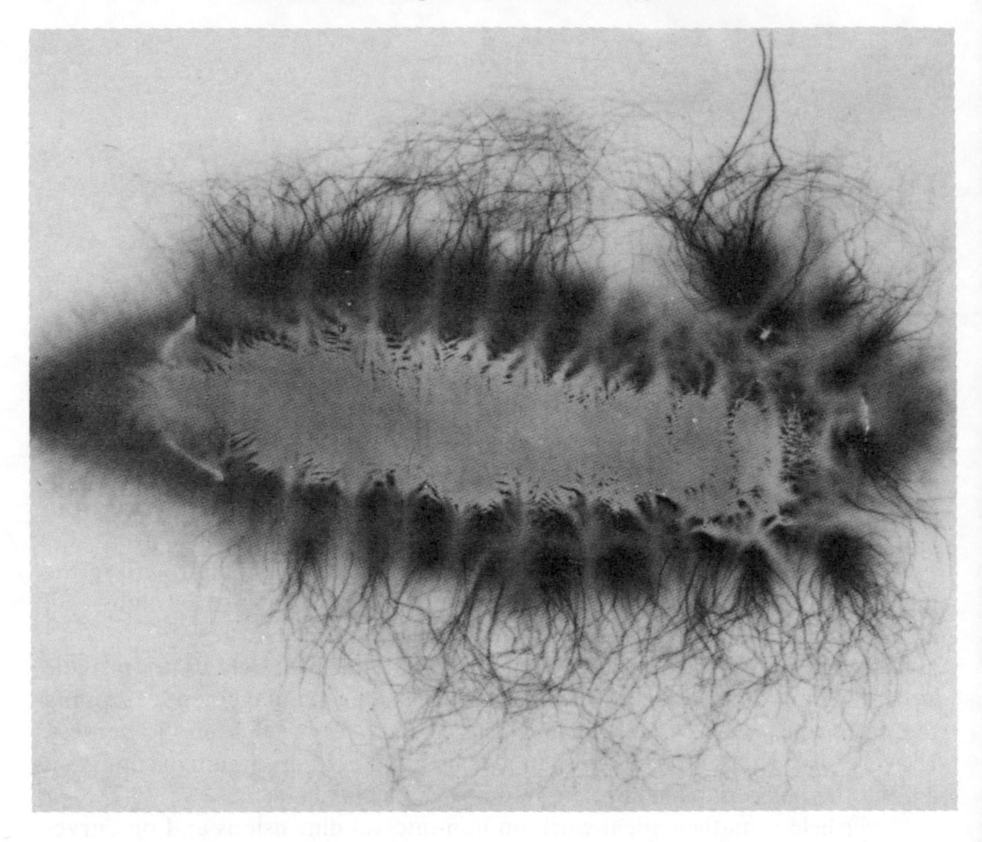

Figure 1.1 Periodic, fractal and chaotic structures are all to be found on this plate. Effluviograph of a coleopter beetle larva (*Plate*: G. Cherbit)

secretary. I would also like to thank D. Verkhoff for his devoted help in bringing this book about, N. Dahmane for the drawings and Madame C. Ferrante for taking care of the typing.

I will end by parodying the humorist P. Dac:

The Fractal in essence is such that the highest international authorities agree in acknowledging it to be the most startling discovery of our time, a discovery which, in a future all the nearer for being less far away, is not only pressed into the service of everything, to say the least, but also of anything, including everything which flows from it, without prejudice to anything else, etcetera, etcetera.

Chapter 2 Sundry observations

Benoît Mandelbrot
Department of Mathematics, Harvard University

I am delighted and moved to have the opportunity to participate in your seminar, for you know that it deals with subjects that are very dear to me. If I am touched, it is because it is hard for me to forget that at one time, not so long ago, I was the only 'fractalist' in the business. Since then, we have multiplied and have managed to get to the next stage, the first conference on fractals, several of whose participants are present here. It was held at Courchevel, in Savoy, in July 1982, and many of those present still wondered if the study of fractals would ever become a 'normal' scientific subject and not merely the dream or lucubrations of one person. Today the subject is well established and has taken firm root, which pleases me greatly, as I am sure you understand, and the monthly seminar of the Membrane Biophysics Research Group is proof of this. Another proof. in 1985, there will be about 10 meetings world-wide on the subject of fractals.

I do not have much of anything organized to say, except perhaps on the Hausdorff-fractal relationship, which has suddenly become interesting. Before getting to the heart of the matter, however, I would like to make various remarks. First of all, I would like to reply to several questions that have often been put to me in private.

Where do I work at the moment and what does my letterhead mean? I have two positions, one at IBM and the other at Harvard. At IBM I am known as an 'IBM Fellow' with a capital letter. The capital letter is essential because, in English, the word fellow has many different meanings. At IBM, the title of Fellow gives me the freedom of a sort of long-term scholarship. At Harvard, I am at the Department of Mathematics as 'Professor of the Practice'. This is an old and extremely vague title, which is good, because Harvard has been able to use it to give me a part-time job. Thus, to translate word for word, making me a 'Professor of Applied Mathematics' would be tempting and would make some sense, but is nevertheless still something of a mistranslation. My main activity, in fact, is the writing of a mathematical textbook on fractals which will include a good proportion of new things.

I am just as often asked questions about the 'Barnard Medal for Meritorious Service to Science' bestowed on me by the National Academy of Sciences of the USA (Washington) in May 1985 for my work on fractal geometry. This medal is given every five years and I confess that the list of my 19 predecessors fills me with awe. Further, I truly felt that my life had reached a turning point on hearing it suggested that I congratulate myself on having known how to rediscover the classical style of natural philosophy. Regarding this, it is a shame that this term is less common in French than in English, for it expresses very well a point of view that has always informed my work but which has previously been described in less apt terms. For example, I have often heard myself accused of putting off the choice of a definite direction in my work and sticking to it. Without having changed, the chronic indecisive, oblivious to boundaries between disciplines, now qualifies as a researcher in natural philosophy!

These light-hearted reminiscences perhaps make you ask which route I took to become a natural philosopher? For a detailed account, I would rather, with your permission, refer you to my contribution to a book entitled *Mathematical People*. This is a collection of interviews and profiles, published in May 1985 by Birkhauser-Boston. In 'inviting' 24 people to appear in this book, the editors have not claimed to have chosen the best 24 mathematicians in the world. The fact that they did not seek anonymity and that Volume II is on the way testifies to this. They were looking for elaborate, surprising and even unbelievable stories. My own career had for a long time been quite chaotic until, several years ago, it seems to have undergone a sudden transition to orderliness. Moreover, it possibly seems this way, with hindsight, because the theme of 'the study of order in chaos' is in the wind and its new meaning applies exactly to what has always been at the centre of my scientific deliberations. In the past, the study of order in chaos was just one definition of science!

My uncle (1899–1983), a mathematician of the College of France and of the Institute, played an important role in my life. He liked to remind me that when I was 20 years old, I told anyone who would listen that my ambition was to find a corner of science, not necessarily very extensive or even significant, of which I would know enough to be its Kepler or even its Newton. This ambition excluded everything that had already been worked out and caused me, on the contrary, to lavish a great deal of attention on those topics best described as 'orphans'. I was even willing, unlike most people, to 'rummage in the dustbins of science'. I mean 'dustbin' quite literally, because two of the main stimuli to my work came from reading items that were about to sink without trace.

The first time, in 1951, I had called in on my uncle for a chat, as I often did, before leaving the Latin Quarter. When I was about to leave, he took from his wastepaper basket a book review that Joseph Walsh had sent him for me to read on the Metro. Walsh, a Harvard mathematician, sent all his publications to my uncle, and amongst the serious articles was the review that he had done of a book by George Kingsley Zipf. Zipf had become the spokesman of a certain idea and he had written a book to tell the world about it. I later learnt that some who knew him, or who had flipped all through his book, thought that Zipf was raving mad. However, Walsh was his friend and he only mentioned things that showed Zipf in a good light. Zipf talked a lot about a certain empirical regularity relating to the frequency of words occurring in speech, a subject of no real interest to anyone, and then of several empirical regularities relating to the economy, which although known to economists were not regarded as interesting by them. To start with, Walsh's review of Zipf's book drew me to the problem of the frequency distribution of words. Even before the end of this memorable Metro journey (my uncle lived some distance from my parents), I had found an explanation for Zipf's law, or rather a much improved generalized version of it. Then Zipf led me on to economics.

The next stage of my work was symbolized by my study of coastlines. This was an important stage, during which my work underwent a profound change, from being purely analytic to having a geometric flavour. Its point of departure was another 'orphan' text rescued from a dustbin, in fact from something of the posthumous work of a great English eccentric, Lewis Fry Richardson. Nowadays, the emphasis would be on 'great' but in those days it would have been on 'eccentric'. He had studied a great variety of things, including turbulence and the solution of differential equations, but all this really amounted to a large collection of minor works. For example, he was interested in the length of the English coastline, among others. Comparing the lengths he measured using various different maps, he noticed that, as the analysis became more and more accurate, the lengths increased in a regular way, which he described by means of an empirical formula. And what importance did Richardson attach to this result? Suffice it to say that he never published it! After his death, a variety of unpublished formulae appeared at once in a survey, which I deemed to be of no immediate interest.

Moreover, some classification was soon going to be needed in mathematics: from among 'the mathematical monstrosities' of the 1900s some, but not others, were going to be 'legitimized' as fractals.

I can already hear the objections. After all, Zipf's report appeared in the *Scientific American*, a pillar of the 'establishment', as opposed to a dustbin, and the journal *General Systems Yearbook* could surely boast more faithful followers than our own work. Very well, I am willing to retract 'dustbins', but I stick to 'abandoned orphans'.

I now come to a third, and more interesting, reason for again talking about the good old days, which is indeed to be found in the title of your seminar. I would like to make it very clear that the Hausdorff dimension did not make an appearance in this business until very much later, and that, for 'political' reasons, I have for a long time taken it upon myself to exaggerate its real importance. Now, at the time of writing this material (autumn 1985), its importance has only diminished. Of course, we all know Dirac's saying, that if a theory is mathematically beautiful and elegant, it is inconceivable that nature does not have a use for it. It may be that this approach has succeeded from time to time, but these are exceptions. Thus, my goal was not above all to find practical applications for those fine examples of reputedly inapplicable things in the general area of Hausdorff dimension. Had I attempted this, I would have undoubtedly got stuck.

Now what about the political reasons to which I alluded a moment ago? For my own practical purposes, I had rediscovered various expressions which, in general, take fractional values and which are of the same type as what I much later named 'fractal dimension'. However, no well-known journal wanted to publish anything on fractal dimension. It was then that, I admit, I engaged in a deliberate piece of intellectual embezzlement. It had to with various notions, already in use in pure mathematics, of which the oldest, and the least obscure, was due to Felix Hausdorff.

In 1954, I was able to get to know Henry MacKean, while he was writing a thesis, at Princeton, on the Hausdorff–Beisicovitch dimension of certain probabilistic sets. I asked myself why this man, so wise and so brilliant, was involved in this narrow and relatively inactive branch of mathematics. (Remember Hermann Weyl's comparison of mathematics with the delta of a river, far from all of whose branches are substantial and active, while, nevertheless, even the slowest has some contribution to make to the mathematical world!) However, whatever my feelings on the subject of MacKean's thesis may have been, I had, thanks to him, a good understanding of the Hausdorff dimension, and, furthermore, my mentor, Paul Lévy, talked about it from time to time. As far as the matter that draws us together here is concerned, the essential feature of the Hausdorff dimension is that it calls for successive passages to limits, which makes it quite useless in science, because it is impossible to measure it. Moreover, it is precisely these passages to the limit that give the Hausdorff dimension the great generality that mathematicians want. Again, since the established journals forbade me to mention, however, briefly, dimensions that had the temerity to be fractions, but of no known pedigree, I 'corrupted' Hausdorff dimension to my own ends. This was a long-established 'vehicle', which had a literature that bore the illustrious names of Hausdorff and Beisicovitch and, more recently, had been significant in harmonic analysis. In particular, there was the work of Raphael Salem and Jean-Pierre Kahane in France, and I was still very much aware of the influence of my uncle, of whom I have already spoken.

Could physics have managed without Hausdorff? We cannot rewrite history and we will never know if this adoption of his work was absolutely necessary. However, mathematician friends soon protested, pointing out that this dimension only dealt with local structures, about which it would be surprising if physics concerned itself. In fact, I can reply to them that their objection was groundless in the case of the first fractals, which turned out to possess an internal similarity, and hence were such that the same tools sufficed for both local and global structures as well as everything in between. Nevertheless, I am reminded of these old criticisms now that I am writing this material (autumn 1985), because, likewise, I am working with fractals with internal similarity. In this context, it is necessary to allow the proliferation of distinct fractal dimensions beyond anything already familiar. On the other hand, however, the use of the Hausdorff dimension becomes controversial, as it has very serious limitations. This was quite unforeseen in earlier times, but, all in all, it is no bad thing that it never really played a central role in the study of fractals.

However, one can never get away with anything, and I, indeed all of us, now have to pay the price for this past tactical manoeuvre. To begin with, the price to be paid depends on the country and context in which one is working, as well as other political cricumstances. Sometimes, this is just being reminded, somewhat embarrassingly, of a certain definition of fractals that I once gave.

Now I think I sometimes hear it said that fractal geometry can only be 'applied' to already existing mathematics or, even worse, that it is no more than an alluring 'buzz-word' appearing on the cover of a book of beautiful illustrations of things already known to specialists. Such a book would, essentially, be no more than a popularization, and two anonymous accounts of my books go so far as to assert that this is precisely how I see my objective! This is all nonsense. Take, for example, the recent treatise of K. Falconer. The word fractal appears in the title, but the contents confirm just how little fractal geometry owes to the mathematics of this century and nothing whatsoever to its mannerisms. This helps to explain the success of fractals, but not much of their substance, which leaves plenty of work for the fractalists. Fractals are not so much the daughter of this century's mathematics but rather their sister, or even their cousin, because they are as beautiful, and just as much the legitimate issue of the inspired constructions of Cantor, Peano, Koch Lebesgue and Sierpinski. There is no rush to generalize these constructions, but rather to study them in detail for what they are and to encourage them to proliferate, something they do with admirable enthusiasm.

It is clear from what I have just said that cousinly relations have become somewhat strained, and this has recently become rather more than just an amusing novelty. What is the source of this tension and how is it that the same seeds from 1900 have produced such different plants? It is surely due to the way experimental mathematics has suddenly burst into life and prospered, thanks to the computer. It seems clear that in ancient times, axiomatic and rigorous mathematics came after the experimental mathematics of those depicted by the artists of the day playing with pebbles and drawing pictures in the sand. Experiment again played a central role in the work of the geometers of just over a century ago, for example Gauss.

However, experimental mathematics then ran out of steam and stopped playing its traditional nutritive role. It is at this point, without doubt, that one should look for the explanation for the importance assumed by a certain well-known historical development, associated with the names of G. H. Hardy and the mythical Nicolas Bourbaki, which led to the systemization and long-time enthronement of the idea that there is a pure mathematics of virgin heredity that reproduces itself by parthenogenesis without any outside contribution. This idea even took root in my uncle, this master of complex analysis, who had been one of the founders of Bourbaki but had left them before their worst excesses. For example, my uncle was never able to convince himself that physics really had been the source of the inspiration for the work in harmonic analysis of his great friend Norbert Wiener.

One of the pleasures of my life has thus been to provide a counter-example, in the form of fractal geometry, to the notion that the best mathematicians can only get really enthusiastic about problems that come entirely from within mathematics itself. This is far from being the only high point of my life, although it is the only one that anyone, not necessarily an expert, can appreciate. Most of the other counter-examples come from modern mathematical physics, with its formal diagrams and formulae, and are so technically difficult as to be beyond the appreciation of any but the specialists.

The tool of fractal geometry is therefore the computer, while I am its workman. It might perhaps be said that the tool was the important thing and that any other worker would have used it just as effectively in the slaying of the Hardy–Bourbaki dragon without the flamboyance that some people accuse me of and find superfluous or even a handicap? Let it be said once more that history cannot be relived, and so there is no point wasting time on such futile discussion. However, I would like to say that the only way of telling whether or not something has become indispensable is to observe the crowds of people lining up to use it. A very familiar phenomenon throughout the history of science is that, as soon as 'the world is ready for it', one sees the same things being discovered simultaneously by a broad range of people who are not even known to one another. Now this phenomenon has been absent in the case of fractal geometry; it only became apparent in 1982, and suddently became general in 1985.

In the light of what has just been said,, it is clear that my real contribution to fractal geometry lies outside anything that can be easily classified. No one would describe me as a creator from scratch, because I am too much of a philosopher and a historian and quote from too many old and mysterious sources. However, this does not prevent there being a breathtaking range of opinions from almost nothing to almost everything, and I feel an urge to sum myself up in a few words. I would therefore say that, if one includes the things that, one way or another, I have brought about, my own feeling is that, up to the moment when fractals suddenly 'took off', I did almost everything.

To be precise, I never had a team of coworkers: it seems that this was the price I had to pay for my exceptional freedom of manoeuvre. I did have a programmer, as well as friends and visitors, all very able and obliging, to help me distil something from all the weird stuff floating around the laboratory. For example, it happened, by complete chance, that there was in the laboratory the prototype of a typographical machine to do with a project that had been abandoned. In this machine, each letter was composed of a collection of very fine (3/100 mm) dots. Several of us at once saw the possibility in this of making up new alphabets. For example, we added an alphabet of 64 letters formed from 8×8 squares of dots, so the squares were of side 2.5 mm, and square number k comprised k dots arranged in as regular a way as possible. Suitably programmed, the machine thought it was printing a long document without any spaces between either the words or the lines. However, it was actually providing us with an image, plotted in advance, involving 64 shades of grey. A single design was equivalent to millions of words! Furthermore, my first book on fractals still contained several designs drawn by a draughtsman. However, the ones that were supposed to be non-random were a little random, while when the draughtsman had to redraw lines in random directions, he had a tendency, easy to understand, to make their directions either nearly vertical or nearly horizontal. Before long, the computer took over all my designs as well as the typesetting, not for reasons of economy but to give me more control over things. In this way, technological changes have enabled control of the text and illustration of a book to pass from the editor to the author. I am not sure that many authors would wish or would even be able to benefit from this. Some would consider it a distraction, but not me. Helped by a few IBM colleagues and motivated by the challenge of it, I have taken complete control of this work, from start to finish, and a most exciting venture it has been.

Bibliography

Falconer K. J. (1985) *The Geometry of Fractal Sets*, Cambridge University Press, Cambridge.

Fischer P. and Smith W (Eds.) (1985) *Chaos, Fractals and Dynamics*, Marcel Dekker, New York.

Mandelbrojt S. (1985) Souvenirs à bâtons rompus, Publications du Séminaire d'Histoire des Mathématiques de l'Université de Paris VI.

Mandelbrot B. B. (1982) *The Fractal Geometry of Nature*, W. H. Freeman (Japanese translation, 1984; German translation, in preparation).

Mandelbrot B. B. (1983) In *Le Débat* (March 1983).

Mandelbrot B. B. (1984a) In *Omni* (February 1984).

Mandelbrot B. B. (1984b) In *Sciences et Techniques* (May 1984).

Mandelbrot B. B. (1984c) *Les Objects Fractals*, 2nd ed., Flammarion (Italian, Spanish and Hungarian translations in preparation).

Mandelbrot B. B. (1985a) In D. Albert and G. I. Alexanderson (Eds.), *Mathematical people*, Birkhauser-Boston.

Mandelbrot B. B. (1986a) In H. O. Peitgen and P. H. Richter (Eds.), *The Beauty of Fractals*, Springer-Verlag.

Mandelbrot B. B. (1986b) In *Interdisciplinary Science Reviews.*

Mandelbrot B. B. *et al.* (1984) *Fractal Aspects of Materials*, Abstracts, Materials Research Society.

Peitgen H. O. and Richter P. H. (1986) *The Beauty of Fractals*, Springer-Verlag.

Pietronero L. and Tozzati M (Eds.) (1986) *Fractals in Physics* (Trieste, 1985), North-Holland.

Shlesinger M. F., Mandelbrot B. B. and Rubin (Eds.) (1984) Fractals in physics (Gaithersburg, 1983), *J. Statist. Phys.*, **36** (5/6).

Stanley H. E. and Ostrowsky N. (1985) *On Growth and Form: Fractal and Non-fractal Shapes*, Martinus-Nijhoff.

Stewart I. (1983) *Fractals*, E. Belin.

Chapter 3 Models of irregular curves

Serge Dubuc
Department of Mathematics University of Montreal

What do we mean by irregular curves? A function will be said to be irregular if its derivative is almost everywhere non-existent. We confine attention to plane curves given by two continuous parametric functions $x(t)$, $y(t)$; we shall say that the curve is irregular if at least one of the two parametric functions is irregular.

I recall some words penned by Henri Poincaré [42] in 1889 in the first volume of the *Journal of Mathematics Education* when he was discussing 'Logic and intuition in the mathematical sciences and in education'.

> There then loomed up a whole mass of strange functions which seemed to be doing their best to resemble the good honest functions is general use as little as possible. No longer continuous, or, if continuous, not possessing derivatives, etc. Moreover, from the point of view of logic, it is these peculiar functions which are actually the more general; indeed, those that one comes across without having looked for them, and which obey simple laws, only appear as a very special case; they occupy but a very small corner of the mathematical world.
>
> In the past, when a new function was invented, it was with some specific practical purpose in view; today they are deliberately invented for the one and only purpose of refuting the reasoning of our predecessors.

Today, it is easier to challenge Poincaŕe's opinion that irregular functions are useless. Brownian motion provides the strongest reason that one can invoke. In 1840 the botanist Brown noticed the disordered nature of the movements of a suspension in a liquid. In 1900 Louis Bachelier [2] saw the close connection between brownian motion and the heat equation and, above all, Einstein [17] in 1905 endorsed the role of brownian motion in physics. A little later, with the help of very detailed photographs, the physicist Jean Perrin [41] emphasized the link between the trajectories of brownian motion and the mathematicians' continuous non-differentiable functions. Perrin's intuition was subsequently confirmed by Wiener, particularly in the joint article [40] with Paley and Zygmund.

I here append a rough sketch of brownian motion (Figure 3.1), in which the irregularity of the trajectory is quite clear. A natural question is how do we measure this irregularity? One possibility is to use the Hausdorff measure introduced by Hausdorff in [22]. Paul Lévy showed in [28] that the Hausdorff

Figure 3.1 Brownian motion

dimension of the trajectory $\{B(t)\colon t\varepsilon[0,1]\}$ of brownian motion is exactly equal to 2 in spite of the fact that the area occupied by the trajectory is a null set (cf. [27] or [29]).

In this article, I shall present various classes of irregular curves and I hope to show you that these curves are of some mathematical interest and that they also have some scientific use, despite Poincaré's scepticism. I shall give some attention to the notion of dimension of a curve, which enables us to quantify the degree of irregularity of a curve. In certain situations, the Hausdorff dimension turns out to be inadequate from either the theoretical or the numerical point of view. I shall recall some remarkable work of Georges Bouligand, done in the late 1920s, and indicate its present significance. I shall also issue a general invitation to again take up a notion due to Lévy, namely the variation of fractional order of a function. I hope to present evidence to show that there is still some point in the mathematical study of dimension. I shall conclude by listing several recent uses of irregular curves in areas outside mathematics.

3.1 Examples of irregular curves

Before considering applications, it is a good idea to have a selection of irregular curves at one's disposal. Since Weierstrass' classic example, a whole host of curves has been dreamed up, and this activity is still going on today. Two of my fellow conferees, Michel Dekking and Michel Mendès-France, are involved in this: see references [8–12] and [38]. For several years, I have been interested in the production of graphs of tangentless curves. I offer you some of the curves that are to be found in [14–16] as well as some new ones. From among them, I have selected those in which the growth of a form is particularly striking.

3.1.1 Curves associated with trigonometric series

To start with, consider the complex-valued function given by the series $\sum_{n=0}^{\infty} a^n e^{i\omega^n t}$ where $|a| < 1$ and ω is real. The real part of $z(t)$ is the Weierstrass function [52], while the imaginary part is related to the function of Cellérier [6]. Figure 3.2 shows, as a diagram in the complex plane, the curve $t \to z(t)$ when a is chosen to be $\frac{1}{2}$ and ω is 2. As the first three terms of the series give a cardioid, this curve is a perturbed cardioid. Hardy [21] showed that both the

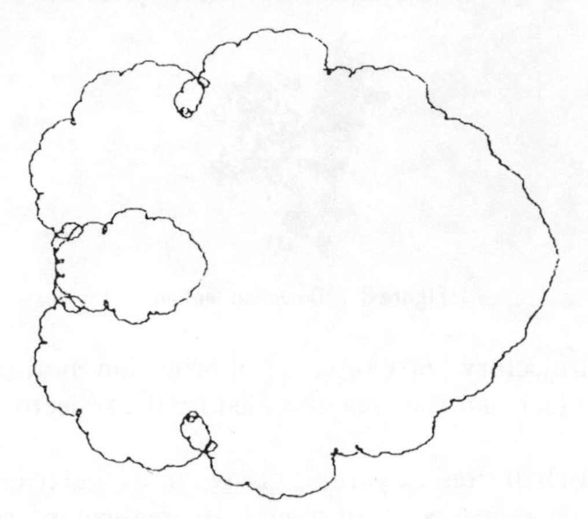

Figure 3.2 Weierstrass and Cellérier functions

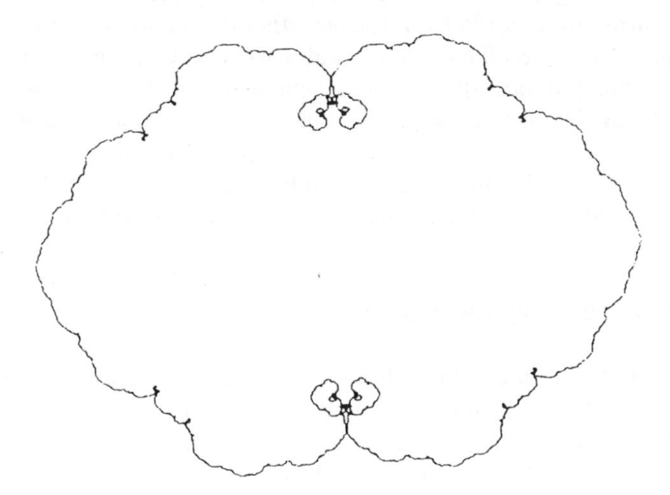

Figure 3.3 Weierstrass and Cellérier functions

real part and the imaginary part of $z(t)$ are functions without derivative. Figure 3.3 shows the same type of curve, this time with $a = \frac{1}{3}$ and $\omega = 3$.

3.1.2 Von Koch–Mandelbrot curves

Benoît Mandelbrot [33] has extended a geometrical construction due to von Koch [51] and has suggested that plane curves may be constructed using a self-similarity property. One such curve is formed from m arcs similar to the curve itself. Let $P_0, P_1, \ldots, P_m$, be $m+1$ points in the euclidean plane $\mathbf{R}^2$, and let S_k denote the similarity of the plane that maps the line segment $[P_0, P_m]$ onto the

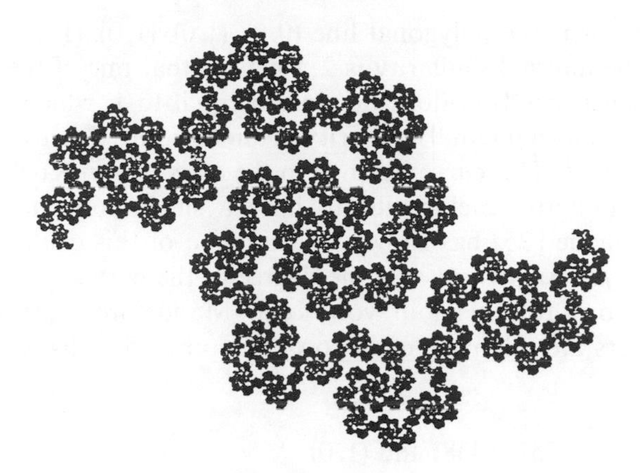

Figure 3.4 Polymer

segment $[P_{k-1}, P_k]$. The von Koch–Mandelbrot curve $\{x(t), y(t)\}, t \in [0, 1]$ is such that $x((k-1+t)/m), y((k-1+t)/m) = S_k(x(t), y(t))$ for each t in $[0, 1]$ and for $k = 1, 2, \ldots, m$. Specifying the points $\{P_k\}_{k=0}^{m}$ suffices to determine the curve. This continuous curve will be well defined if each of the ratios r_k of the similarities S_k is less than 1, where r_k is the quotient $\|P_k - P_{k-1}\| / \|P_m - P_0\|$. The construction of this curve can be realized as a uniform limit of polygonal lines according to von Koch's original method. We shall refer to m as the degree of the curve. The irregularity of the curve is adequately described by the dimension of the internal similarity, which is the solution d of the equation $r_1^d + r_2^d + \cdots + r_m^d = 1$. The further the number d is from 1, the more pronounced is the irregularity.

Figure 3.4 resembles a polymer. It is the von Koch–Mandelbrot curve of degree 3 whose generator is formed from the four points: (0, 0), (0.250, 0.297), (0.750, −0.297), (1, 0). The dimension of the internal similarity is 1.800.

Figure 3.5, called 'forest' in [14], is the von Koch–Mandelbrot curve of degree

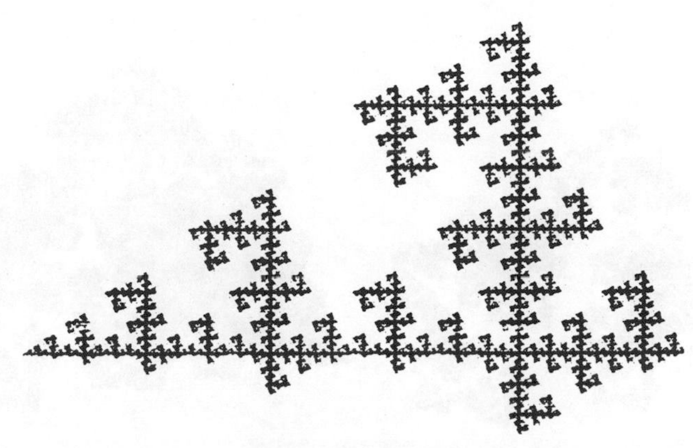

Figure 3.5 Forest

4 whose generator is the polygonal line $(0,0)$, $(\frac{1}{2},0)$, $(1,0)$, $(1,\frac{1}{2})$ and $(1,0)$. The dimension of the internal similarity is 2. I believe that one of the properties of this curve is that it has Hausdorff dimension equal to 1, which would be surprising for a curve with internal similarity of dimension 2. Each similarity coefficient has the value $\frac{1}{2}$. This curve belongs to the class of von Koch–Mandelbrot curves whose similarity coefficients all have a single constant value p, say. Jean–Pierre Kahane [25] has shown that a curve of this class is Lipschitz of order $\lambda = -\log p/\log m$, where m is the degree of the curve.

Figure 3.6 shows together four von Koch–Mandelbrot curves of degree 3 whose generators are relatively close to one another. The list of generators is as follows:

$(0,0)$, $(0.503,0)$, $(0.752,0.438)$ and $(1,0)$
$(0,0)$, $(0.543,0)$, $(0.772,0.493)$ and $(1,0)$
$(0,0)$, $(0.577,0)$, $(0.788,0.537)$ and $(1,0)$
$(0,0)$, $(0.607,0)$, $(0.803,0.574)$ and $(1,0)$

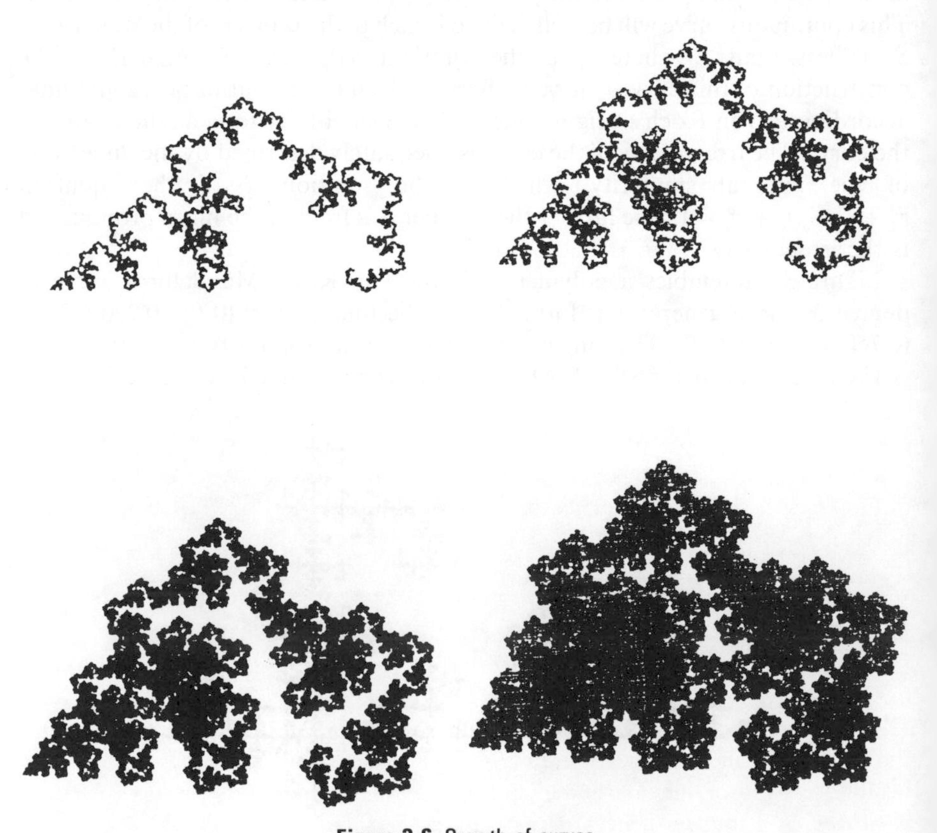

Figure 3.6 Growth of curves

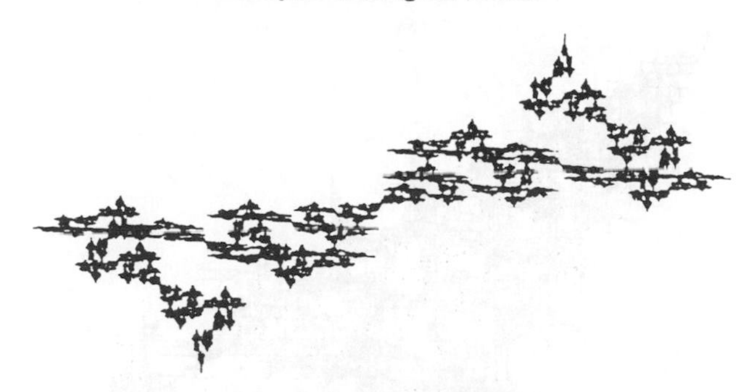

Figure 3.7 A hyperbolic curve

The dimensions of internal similarity are 1.6, 1.8, 2 and 2.2 respectively. It comes as no surprise to see that a curve admits more and more multiple points as its dimension parameter increases.

3.1.3 Hyperbolic curves

A degenerate case of the von Koch–Mandelbrot curves that is worth looking at occurs when the polygonal generator is part of a single straight line. To fix ideas, consider the case where the generator lies on the x axis and is given by the points $(0,0)$, $(x_1, 0), \ldots, (x_{m-1}, 0)$ and $(1,0)$. Put $x_0 = 0$ and $x_m = 1$. The associated von Koch–Manderbrot curve is then of the form $(x(t), 0)$, where $x(t)$ is the bounded function such that $x((k-1+t)/m) = x_{k-1} + (x_k - x_{k-1})x(t)$ for each t in $[0,1]$ and for $k = 1, 2, \ldots, m$. Giving the points $\{x\}_{k=0}^{m}$ suffices to determine the function $x(t)$. This continuous function will be well defined if each of the numbers $|x_k - x_{k-1}|$ is less than 1.

Now one can give a second generator lying on the y axis and given by the points $(0,0)$, $(0, y_1), \ldots, (0, y_{m-1})$ and $(0,1)$. Put both y_0 and y_m equal to 1. A second von Koch–Mandelbrot curve of the form $(0, y(t))$ is thus constructed; $y(t)$ is the bounded function such that $y((k-1+t)/m) = y_{k-1} + (y_k - y_{k-1})y(t)$ for each t in $[0,1]$ and for $k = 1, 2, \ldots, m$. The hypothesis is still that each of the numbers $|y_k - y_{k-1}|$ is less than 1.

The curve $(x(t), y(t))$ thus obtained has been given the name 'hyperbolic curve' in [14] (see Figure 3.7). The generator of a typical hyperbolic curve will be the polygonal line formed from the points $(x_0, y_0), (x_1, y_1), \ldots, (x_{m-1}, y_{m-1})$ and (x_m, y_m).

The next three figures show hyperbolic curves. The first of them is of degree 5 and its generator is given by $(0,0)$, $(\frac{1}{2}, -\frac{1}{4}), (\frac{1}{4}, \frac{1}{4}), (\frac{3}{4}, \frac{3}{4}), (\frac{1}{2}, \frac{5}{4})$ and $(1,1)$.

The second is also of degree 5, has the look of a fir tree (Figure 3.8) and has for its generator the points $(0,0)$, $(-\frac{1}{5}, -\frac{1}{5})$, $(-\frac{2}{5}, \frac{3}{5}), (\frac{2}{5}, \frac{7}{5}), (\frac{6}{5}, \frac{6}{5})$ and $(1,1)$.

Being in Paris, I cannot resist showing you a curve revealed in [14], namely the 'Russian' Eiffel tower (Figure 3.9); like Russian dolls, it repeats itself within itself floor by floor. The parameters of this curve are extremely simple: the curve is of degree 3 and its generator is $(0,0)$, $(-\frac{1}{3}, \frac{2}{3})$, $(\frac{1}{3}, \frac{4}{3}), (1,1)$. The source of the

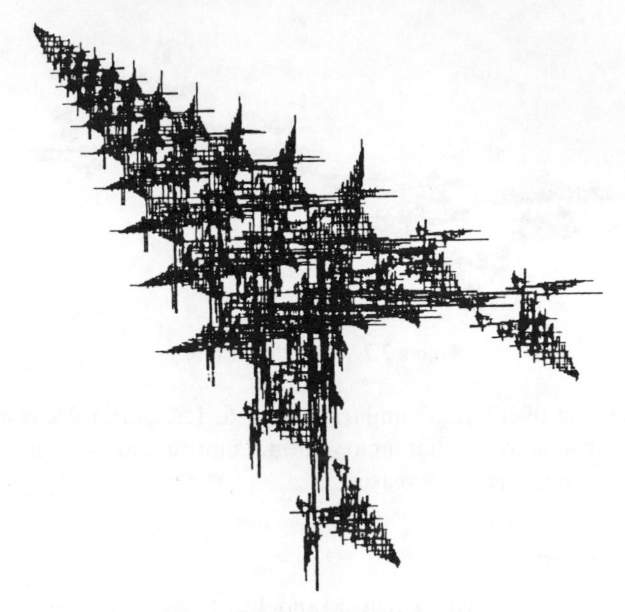

Figure 3.8 The hyperbolic fir tree

Figure 3.9 'Russian' Eiffel tower

Russian doll effect is the vector from the second to the third point of the generator; this vector is $(\frac{2}{3}, \frac{2}{3})$, a multiple of the vector joining the endpoints of the generator.

3.2 Curves modelled by a functional equation

Each of the curves shown in Figures 2 to 9 above satisfied a functional equation of the same type and this equation allows one to construct various curves, both new and familiar. I illustrated this in [15] and [16], and briefly review this work here. The functional equation with which we are concerned is

$$x(t) = f(t, x(b(t))) \tag{1}$$

where t varies in T, x varies in X, $f(t, x)$ is a function of two variables t and x with values in X and $b(t)$ is a mapping of T into itself. The unknown function is the function $x(t)$ from T into X. To our knowledge, Read [46] was the first person to introduce the equation (1) in the setting where both T and X were the complex plane and b and f were analytic functions; he looked for an analytic solution to equation (1). In 1957, Bajraktarevic [3] widened the scope of this work by proposing the study of the functional equation

$$x(t) = f(t, x(b_1(t)), x(b_2(t)), \ldots, x(b_r(t))) \tag{2}$$

where T and X are subsets of two euclidean spaces, t varies in T, the functions $\{b_i\}_{i=1}^{r}$ are transformations of T and the function f is a mapping of $T \times X^r$ into X. Equation (2) has since attracted the attention of a number of mathematicians, several Polish scholars among them.

Our interest, however, is confined to the case where $r = 1$, and we return to the functional equation (1). Let us recapitulate the essence of Bajraktarevic's argument. T is an arbitrary set, t varies in T and $b(t)$ is a transformation of T into itself. X is a complete metric space whose metric will be denoted by d. $f(t, x)$ is a mapping of $T \times X$ into X. Let us define two conditions that f may satisfy:

(a) There exists a point x_0 of X such that $\{f(t, x_0): t \in T\}$ is a bounded subset of X.
(b) There exists a number $k < 1$ such that for each t of T and for each pair of points of X, x_1 and x_2, $d(f(t, x_1), f(t, x_2)) \leqslant kd(x_1, x_2)$.

Theorem 1 When conditions (a) and (b) are satisfied, equation (1) admits one and only one bounded solution. If $x_0(t)$ is a bounded function from T into X and if $x_n(t)$ is defined recursively by

$$x_{n+1}(t) = f(t, x_n(b(t))), \qquad \text{for } n = 0, 1, 2, \ldots,$$

then the solution of equation (1) is the limit of the sequence $(x_n(t))$, in the sense of uniform convergence over T.

Note that if T is a topological space, and if f and b are continuous functions, the solution $x(t)$ is a continuous function. Here is a sufficient condition for the solution $x(t)$ of equation (1) to be a continuous function when T is a topological space and the functions b and f are not necessarily continuous.

(c) There exists a family F of continuous functions from T into X such that, for each $y(t)$ of F, $f(t, y(b(t)))$ belongs to the family F.

We now use the von Koch–Mandelbrot curves to illustrate the functional equation (1). Let $P_0, P_1, \ldots, P_m$ be points of the euclidean plane $\mathbf{R} \times \mathbf{R}$, and let S_k denote the similarity of the plane that maps the line segment $[P_0, P_m]$ onto the segment $[P_{k-1}, P_k]$. For T, we take $[0, 1]$; $b(t)$ is the fractional part of m times t. For X, we take the plane $\mathbf{R}^2$. If (x, y) is a point of X, $f(t, x, y)$ is defined to be $S_{k+1}(x, y)$, where k is the integer part of m times t. If each of the ratios r_k of the similarities S_k is less than 1, condition (b) is satisfied. Conditions (a) and (c) are easily seen to be satisfied. The construction of the von Koch–Mandelbrot curve is thus a special case of Theorem 1.

The scope of Theorem 1 is limited by condition (b). However, Baron [4] has weakened this condition. Finally, I should say that the first mathematician to deserve the credit for using functional equations to construct irregular curves is Georges de Rham, as one can see in [13]. This approach is very fertile. There is another distinct, but equally fruitful, approach, namely the use of a theory of substitutions as to be found in Dekking [8].

3.3 Notions of dimension

Starting with an irregular continuous curve parametrized by means of two functions $x(t)$ and $y(t)$ with the parameter t varying in a set T, generally an interval, the notion of size can be considered from the point of view of a fractional dimension exceeding one. We must first identify our objectives. Are we just going to try to calculate the size of the trace of the curve in the plane, i.e. the set $C = \{(x(t), y(t)) \colon t \in T\}$, or are we to take account of the multiple points of the curve and the way in which C is described? To begin with, we restrict ourselves to the first objective of measuring the size of C. Since C is a compact subset of the plane, the size of C can be obtained using the fractional measure of Hausdorff. In order to make the terminology precise, we here recall Hausdorff's definitions.

3.3.1 Hausdorff dimension

Let (X, d) be a metric space. If A is a subset of X, the diameter of A is the number $\delta(A) = \sup\{d(x, y) \colon x \in A \text{ and } y \in A\}$. For a given real number p and an $\varepsilon > 0$, we write $m_p^\varepsilon(X)$ for the infimum of the set of all numbers of the form $\sum_{k=1}^{\infty} [\delta(A_k)]^p$ where $A_1 \cup A_2 \cup \cdots = X$ is a countable partition of X into subsets A_k of diameter $\delta(A_k)$ less than ε.

The Hausdorff measure of order p of X, $m_p(X)$, is the supremum of the quantities $m_p^\varepsilon(X)$ as ε varies over all positive values.

The Hausdorff dimension of X is the infimum of the set of those p such that $m_p(X) = 0$.

3.3.2 Dimension of internal similarity

To carry out the calculations associated with the Hausdorff dimension can often be tricky. In such a situation, one looks to other ways of defining dimension. In particular, if X is a compact subset of $\mathbf{R}^n$ with enough internal similarities, one can define a dimension of internal similarity. Suppose, therefore, that X is a compact subset of $\mathbf{R}^n$ for which there exist d similarities $\{S_k\}_{k=1}^d$ such that X is the union of the d images $S_k(X)$. We assume that each of the similarity ratios r_k of the similarities S_k is less than one. The dimension of internal similarity is defined to be the solution p of the equation $r_1^p + r_2^p + \cdots + r_d^p = 1$.

It is easy to check that the dimension of internal similarity is either equal to or greater than the Hausdorff dimension. It is enough to cover X by sets of the form $S_i(S_j \cdots (S_k(X)) \cdots)$ where a large number of the S_k's are composed. There are situations where it is clear that the Hausdorff dimension is not the same as the dimension of internal similarity. For example, if X is a subset of the plane, its Hausdorff dimension is at most equal to two; if the dimension of internal similarity ever exceeds two, the two dimensions are different.

Nevertheless, Hutchinson [24] has found sufficient conditions on X and on the similarities S_k for the two dimensions to be the same. Jacques Marion [34–37] has also discovered sufficient conditions. Unfortunately, these conditions are seldom met in the case of a von Koch–Mandelbrot curve. A natural conjecture is that the Hausdorff dimension of a simple von Koch–Mandelbrot curve is the same as its dimension of internal similarity. In the article [36] and, in greater detail, in the technical report [37], Marion establishes the truth of this conjecture.

3.3.3 Bouligand's dimensional order

In a masterly piece of work, Bouligand, in [5], sets out three equivalent ways of defining the dimension of a non-empty compact subset E of euclidean space $\mathbf{R}^3$. The first method is to involve what he calls the Cantor–Minkowski construction in defining the set E_ρ of points whose distance from E is less than ρ, i.e. all points y of space for which there exists a point x of E such that $\|x - y\| < \rho$. He introduces the function $f(\rho)$ of ρ as the Lebesgue volume of the open set E_ρ. The way $f(\rho)$ behaves as ρ decreases to zero enables one to define the dimensional order of E. The lower dimension of E is the number $3 - \alpha$ where α is the supremum of the numbers p such that the limit of $f(\rho)/\rho^p$ as ρ decreases to 0 is zero.

In the second method, given a radius ρ, Bouligand considers a system S of spheres of radius ρ, which are either non-intersecting or at most touching, and which meet the following two conditions: firstly, the centre of each sphere lies in E and, secondly, every sphere of radius ρ with centre a point of E distinct from the centres of the spheres of the system S meets at least one sphere of the system S. If K_ρ is the number of spheres in the system S, Bouligand shows that the number $K_\rho^3/f(\rho)$ always lies between two fixed positive numbers independent of E, of ρ and of the choice of the particular system S. The volume occupied by the spheres of the systems S thus enables the dimensional order of E to be evaluated.

In the third method, Bouligand considers, for each positive integer N, the covering of space by closed cubes of the form $[i/N, (i+1)/N] \times [j/N, (j+1)/N] \times [k/N, (k+1)/N]$. He calls any cube of the covering which meets E a 'contributory' cube. He introduces the function $\varphi(\rho)$, for $\rho = 1/N$, as the total volume of the contributory cubes. He shows that there exist two positive constants a and b (independent of E) such that, for each ρ of the form $1/N$, $af(\rho) < \varphi(\rho) < bf(\rho)$. The behaviour of the sequence $\varphi(1/N)$ thus enables one to define a dimensional order, which will be equivalent to the first. Bouligand had previously obtained the following inequality for the function f: for every θ between 0 and 1, $\theta^3 f(\rho) < f(\theta\rho) < f(\rho)$. We observe that the dimensional order is even determined by the behaviour of the sequence $\{\varphi(2^{-k})\}_{k=0}^{\infty}$. We shall come back to this later.

3.3.4 The ε-entropy and ε-capacity of Kolmogoroff

Starting from a definition of Pontrjagin–Schnirelmann [43], Kolmogoroff [26] has developed the notion of ε-entropy. If A is a compact subset of a metric space (X, d) and if ε is positive, $N(\varepsilon)$ denotes the smallest number of subsets of X, of diameters not more than ε, needed to cover A. The ε-entropy of A is $H_\varepsilon(A) = \log N(\varepsilon)$. The lower dimension of A is the limit inferior of $H_\varepsilon(A)/-\log(\varepsilon)$. The upper dimension is similarly defined. One can show that if (X, d) is euclidean space, the notion of dimension obtained from the ε-entropy is equivalent to the notion of Bouligand, because the number of 'contributory cubes' in a paving of side ε is always of the order of $N(\varepsilon)$.

If A is a compact subset of X and if ε is positive, $M(\varepsilon)$ denotes the largest number of points that one can choose in A in such a way that the distance between any two of them exceeds ε. The ε-capacity of A is $M_\varepsilon(A) = \log M(\varepsilon)$. Since the ε-capacity behaves in much the same way as the ε-entropy, the dimension arising from the ε-capacity coincides with the ε-entropy dimension. In $\mathbf{R}^n$, the ε-capacity is related to Bouligand's second construction of dimensional order.

3.3.5 Other notions of dimension

In [49], Claude Tricot has compared notions of dimension associated with a variety of contexts and authors. Elsewhere, in [50] in particular, he has himself introduced a new notion of dimension, namely that known as the 'packing' dimension. The packing dimension is like a synthesis of ε-capacity and Hausdorff dimension.

3.4 Calculating the size of a curve

We return to the matter of measuring the size of an irregular curve. The measure that best lends itself to numerical calculations is Bouligand's method of 'contributory cubes'. Let K be a subset of the plane whose size one wishes to measure and, for each value of the natural number n, consider the mesh of the plane formed by the lines $x = i/n$ and $y = j/n$ where i and j vary through the appropriate set of integers. This mesh produces squares $Q^n_{i,j}$ of side $1/n$, where the lower left-hand corner of the square $Q^n_{i,j}$ is the point $(i/n, j/n)$. According to Bouligand, the square Q^n_{ij} is contributory if it meets K. In the language of information theorists, the point (i, j) is one of the 'pixels' of K. Thus $\varphi(1/n)$ will, by definition, be the total area of the contributory squares, viz. the area of an elementary square, $1/n^2$, multiplied by the number of pixels of K. The dimensional order of K is given by the sequence of numbers $\varphi(1/n)$. Indeed, more can be said.

Theorem 2 The dimensional order of K can be obtained from the sequence $\varphi(2^{-n})$.

We showed this in the preceding section. Moreover, it is fortunate that knowledge of the pixels for calculating $\varphi(2^{-n})$, when n has some specific value, enables one to calculate $\varphi(2^{-k})$ for $k = 0, 1, \ldots, n-1$. When the dimension of K is well defined, it is obtained numerically from the quantity $-\log \varphi(2^{-n})/(n \log 2)$. A linear regression on the points $\{(k, -\log(\varphi(2^{-k})\}^n_{k=0}$ allows one to control, to some extent, the accuracy of the estimate of the dimension. Bouligand's dimension (or Kolmogoroff's, which is equivalent to it) can thus be simply estimated by this method.

We now discuss the problem of measuring the size of an irregular curve when the multiple points of the curve are taken into account. Only way of achieving this is to return to a concept due to Lévy, in defining the variation of order p of a curve. If $f(t)$ is a continuous function defined on the interval $[a, b]$ and taking values in the plane, and if p is a real number greater than or equal to one and ε is a positive number, we denote by V_p the supremum of the set of quantities of the form $\sum^n_{k=1} \| f(t_k) - f(t_{k-1}) \|^p$ where n is a variable natural number and $a = t_0 < t_1 < \cdots < t_n = b$ is a partition of $[a, b]$ of diameter not exceeding ε (i.e. for each $k, t_k - t_{k-1} \leqslant \varepsilon$). We defined V_p to be the limit of V^ε_p as ε decreases to zero. The degree of irregularity will, by definition, be the infimum of the set of those p for which $V_p = 0$. A non-constant, real-valued, continuous function of bounded variation has a degree of irregularity equal to one. A function that is Lipschitz of order r has a degree of irregularity not exceeding $1/r$. In [29] (pp. 18–20), Lévy shows that the degree of irregularity of linear brownian motion is almost surely equal to 2. The next result shows the relevance of the notion of variation of order p in connection with the theory of dimension.

Theorem 3 If $x(t), y(t)$ is a von Koch–Mandelbrot curve and p is its dimension of internal similarity, the degree of irregularity of the curve is exactly equal to p.

Proof Consider a von Koch–Mandelbrot curve of degree m with generator $P_0, P_1, \ldots, P_m$. Let L denote the distance from P_0 to P_m and r_k the quotient $\|P_k - P_{k-1}\|/L$. We should assume that each of the numbers r_k is less than 1. The dimension of internal similarity is the solution of the equation

$$r_1^p + r_2^p + \cdots + r_m^p = 1.$$

In the parametrization $x(t), y(t)$ of the curve, t varies between 0 and 1. Let us check that, for each $\varepsilon > 0$, $V_p^\varepsilon \geqslant L$. To do this, we choose an integer n such that $h = 1/d^n < \varepsilon$. Using the partition of $[0, 1]$ formed by the multiples of h, we obtain

$$\sum_{k=1}^{d^n} [|x(kh) - x((k-1)h)|^2 + |y(kh) - y((k-1)h)|^2]^{p/2} = L(r_1^p + r_2^p + \cdots + r_m^p)^n = L,$$

from which it follows that $V_p^\varepsilon > L$. Hence $V_p > 0$.

Suppose now that $q > p$; we shall show that $V_q = 0$. We first check that V_q is finite. To see this, we introduce the function $f(t)$ defined by $f(t) = (x(t), y(t))$ as well as the auxiliary quantity $V_q(n)$ which is the supremum of the set of numbers of the form $\sum_{k=1}^n \|f(t_k) - f(t_{k-1})\|^q$ over all partitions $a = t_0 < t_1 < \cdots < t_n = b$ of $[a,b]$. The continuity of f ensures that $V_q(n)$ is finite and that there exists a partition $a = t_0 < t_1 < \cdots < t_n = b$ for which $V_q(n) = \sum_{k=1}^n \|f(t_k) - f(t_{k-1})\|^q$. Looking at this sum, we distinguish those segments $(f(t_{k-1}), f(t_k))$ such that one of the numbers i/m is interior to the interval $[t_{k-1}, t_k]$ and rearrange the remaining segments into m polygonal lines. If d is the diameter of the curve, we obtain the inequality: $V_q(n) \leqslant (m-1)d + \sum_{i=1}^m r_1^q V_q(n)$, whence $V_q(n) \leqslant (m-1)d/(1 - \sum_{i=1}^m r_i^q)$. Since it is the limit of the bounded monotone sequence $V_q(n)$, V_q is therefore finite.

To deduce that V_q is actually zero, it is enough to choose a number r lying between p and q. Because V_r is finite and the function $f(t)$ is uniformly continuous, a simple argument now leads to the conclusion that V_q is zero.

Remark We have established that V_p is not zero, but the proof does not show that V_p is a finite number. Whether or not V_p is finite is still an unsettled question.

3.5 Applications

I here indicate some applications arising from the subject of irregular curves. Several applications have already been described in Mandelbrot's works—see [33]. One of Mandelbrot's first attempts to make use of irregular curves had to do with the form of watercourses or geographical boundaries [30]. In due course, he brought in deterministic curves and then stochastic curves [31]. His ideas stirred up controversy among geologists, such as Scheidegger [47], Hâkanson [20] and Goodchild [19], the last of whom appreciated the interest of this application to geology.

A potential application of irregular curves is in the subject of the creation of forms. In the 1950s, the architect André Hermant [23] drew attention to the presence of complex mathematical forms in nature as well as in the man-made world. He gave the following examples: the shape of a flower or of a leaf, letters of the alphabet, the Colorado delta, crystallization and coral, seashells, skeletons of radiolaria, architectural structures.... Simple curves such as spirals or binary trees can serve as models for the shape of plants or of trees. Hermant himself proposed such models; more recently, Aano and Kunii [1] have put forward very good graphical representations of (simplified) trees. It is, however, by no means out of the question that irregular curves best complete the description of a vegetable object.

Another place where irregular curves occur naturally is in large molecular chains such as polymers or proteins. According to Cotton *et al.* [7], three European teams, the French one enlivened by H. Benoît, incidentally, confirmed in the 1970s that polymers are indeed formed from a complete intermingling of molecules. In view of the experiments of Stapleton *et al.* [48], a reasonable assumption is that proteins occupy space whose fractional dimension is about $\frac{5}{3}$. The theoretical description of long molecular chains is complicated by the phenomenon of excluded volume, i.e. two segments must not interesect again within the space. Mandelbrot [32] has suggested a model for polymers. Also, a renewed interest in random walks without loops has arisen, with several physicists embarking on work on such walks. From the work in this area, I mention only the article [45] of Rammal where the statistics of various random walks are discussed.

Stochastic irregular curves also have plenty to contribute in the area of image processing, see Fournier *et al.* [18]. Brownian motion, both standard and fractional, has its place in the toolbox of computer graphics. Realistic representations of stones, clouds and trees, as well as relief maps, make good use of the existence of irregular curves. Even if the object to be represented is a surface, one needs to study the boundary of a projection of this surface, which is a curve.

Joël Quinqueton, with Marc Berthod [44] and with Nguyen [39], has found an unexpected application of Peano curves which fill up the unit square. In order to analyse the texture of an image, they have put forward the idea of a Peano 'balayage', an operation inverse to that of forming the Peano curve. Under this balayage, a shape lying within the square is transformed into a (one-dimensional) signal; the analysis of the image is thus reduced to the analysis of a signal.

Several applications of irregular curves are currently being developed. To end with, I have some comments to make.

(a) Construction of new curves is always possible, especially in the context of some specific research goal.
(b) If there is explosive growth in the production of models of curves, there is, from the theoretical point of view, much to be done in the way of mathematical analysis of them. For example, in the majority of cases, the Hausdorff dimension of the trace of the curve has not been determined.
(c) The comparison of models of curves with experimental findings has not always been pursued very vigorously.

(d) In the field of irregular curves, there is a schism between mathematicians and information theorists. The mathematicians have little knowledge of the information underlying the graphical representations, while those who create the images do not know much about the underlying mathematics.

I would like to thank several people who have had an indirect, but nevertheless important, influence on this work. I thank Paul Brandouy and Gilles Deslauriers whose bibliographical researches have been a most valuable aid on many occasions. I also thank Claude Tricot, whose thesis caused me to discover the works of Bouligand.

References

1. Aano M., and Kunii T. L. (1984) Botanical tree image generation, IEEE CG A, pp. 10–34.
2. Bachelier L. (1900) Théorie de la Spéculation, *Paris Thesis, Ann. Ec. Norm. Sup. s 3*, **17**, 21–86.
3. Bajraktarevic M. (1957) Sur une équation fonctionelle, *Glasnik Mat.-Fiz. I Astr.*, **12**, 201–205.
4. Baron K. (1974) Note on continuous solutions of a functional equation, *Aequationes Math.*, **1974**, 267–269.
5. Bouligand G. (1928) Ensembles impropres et nombre dimensional. *Bull Sci. Math.*, **52**(2), 320–344, 361–376.
6. Cellérier C. (1890) Note sur les principles fondamentaux de l'analyse, *Bull. Sci. Math (2)*, **XIV** Premiere Partie, 142–160.
7. Cotton J. P., Jannick G. and Picot C. (1976) Neutrons, polymeres et transitions de phase, *Recherche*, **7**(64), 163–165.
8. Dekking F. M. (1982) Replicating superfigures and endomorphisms of free groups, *J. Combin. Theory, Series A*, **1982**, 315–320.
9. Dekking F. M. (1983) Variations on Peano, *Nieuw Arch. Wisk.*, **28**, 275–281.
10. Dekking F. M. (1983) Recurrent sets. *Adv. Math.*, **44**(1), 78–104.
11. Dekking F. M. and Mendès-France M. (1981) Uniform distribution modulo one: a geometrical viewpoint, *J. Reine Angew. Math.*, **1981**, 143–153.
12. Dekking F. M., Mendès-France M. and van der Poorten A. (1982) Folds! (I, II, III), *Math. Intelligencer*, **4**, 130–138, 173–181 and 190–195.
13. de Rham G. (1956–7) Sur quelques courbes définies par des équations fonctionelles, *Rend. Sem. Math. Univ. e Politec. Torino*, **16**, 101–113.
14. Dubuc S. (1982) Une foire de courbes sans tangentes, in *Actualités Math., Actes VIème Congrés Mathématiciens d'Expression Latine*, Gauthier-Villars, Paris, pp. 99–123.
15. Dubuc S. (1984) Functional equations connected with peculiar curves, in *Proceedings of Symposium on Interaction Theory*, Lochau, Austria, Springer Lecture Notes, 1985, 12 pages (to appear).
16. Dubuc S. (1985) Une equation fonctionelle pour diverses constructions geometriques, *Ann. Sci. Math. Quebec*, 33 pages (to appear).

17. Einstein A. (1905) Uber die von der molekularkinetischen Theorie der wärme geforderte Bewgung von in ruhenden Flüssigkeiten suspendierten eilchen, *Annalen der Physik ser 4*, **17**, 549–560.

18. Fournier A., Fussell D. and Carpenter L. (1982) Computer rendering of stochastic models, *Comm ACM*, **1982**, 371–384.

19. Goodchild M. F. (1980) Fractals and the accuracy of geographical measures, *Math Geology*, **23**(2), 85–98.

20. Hâkanson L. (1978) The length of closed geomorphic lines, *Math. Geology*, **10**, 141–167.

21. Hardy G. H. (1916) Weierstrass's non-differentiable function *Trans. Am. Math. Soc.*, **17**, 301–325.

22. Hausdorff F. (1919) Dimension und äusseres Mass. *Math. Ann.*, **79**, 157–179.

23. Hermant A. (1959) *Formes Utiles*, Edition du Salon des Arts Menagers, Paris.

24. Hutchinson J. E. (1981) Fractals and self-similarity, *Indian Univ. Math. J.*, **30**, 713–747.

25. Kahane J.-P. (1970) Courbes étranges, ensembles minces, *Bull. APMEP*, **275/276**, 325–339.

26. Kolmogorov A. N. and Tihomiroff V. M. (1959) ε-Entropy and ε-capacity of sets in functional spaces, *Uspehi Mat. Nauk.*, **14**(2), 3–86; (1961) *Am. Math. Soc. Transl.*, **17**(2), 277–364.

27. Lévy P. (1940) Le mouvement brownien plan, *Am. J. Math.*, **62**, 487–550 (*Oeuvres de P. Lévy*, Vol. V, 1980, Gauthier-Villars, Paris, pp. 20–74).

28. Lévy P. (1953) La mesure de Hausdorff de la courbe du mouvement brownien, *Giornale d. Instituto Ital. d. Attuari*, **XVI**, 1–37 (*Oeuvres de Paul Lévy*, Vol. V, 1980, Gauthier-Villars, Paris, pp. 94–130).

29. Le'vy P. (1954) Le mouvement brownien, in *Mémorial des Sciences Mathématiques*, Gauthier-Villars, Paris (*Oeuvres de Paul Lévy*, Vol. V, 1980, Gauthier-Villars, Paris, pp. 138–218).

30. Mandelbrot B. B. (1967) How long is the coast of Britain? Statistical self-similarity and fractional dimension, *Science*, **155**, 636–638.

31. Mandelbrot B. B. (1975) Stochastic models for Earth relief, the shape and the fractal dimension of the coastlines, and the number-area rule for islands, *Proc. Nat. Acad. Sci., USA*, **72**(10), 3825–3828.

32. Mandelbrot B. B. (1978) Colliers aléatoires et une alternative aux promenades au hasard sans boucle: les cordonnets discrets et fractals, *C.R. Acad. Sci., Paris*, **286**, 933–936.

33. Mandelbrot B. B. (1982) *The Fractal Geometry of Nature*, W. H. Freeman, San Francisco.

34. Marion J. (1979) Le calcul de la mesure de Hausdorff des sous-ensembles parfaits isotypiques de $\mathbf{R}^m$, *C.R. Acad. Sci., Paris*, **289**, Série A, 65–68.

35. Marion J. (1981) Sur la mesure de Hausdorff des ensembles parfaits de translation et de certains ensembles analogues, *C.R. Acad. Sci., Paris*, **292**, Série A, 263–266.

36. Marion J. (1984) Dimension de Hausdorff d'une courbe circulaire simple de von Koch, *C.R. Math. Rep. Acad. Sci. Canada*, **VI**(1), 21–24.

37. Marion J. (1985) Mesures de Hausdorff d'un fractal a similitude interne, Rapport Technique, Dept. de Math., Univ. de Montreal.

38. Mendès-France M. (1984) Folding paper and thermodynamics, *Phys. Rep.*, **103**(1–4), 161–172.
39. Nguyen P. T. and Quinqueton J. (1982) Space filling curves and texture analysis, IEEE, pp. 282–285.
40. Paley R. E. A. C., Wiener N. and Zygmund A. (1933) Notes on random functions, *Math. Z.*, **37**, 647–668.
41. Perrin J. (1913) *Atomes*, Gallimard, Paris (Reprinted 1970).
42. Poincaré H. (1889) La logique et l'intuition dans la science mathématique et dans l'enseignement. *Enseignement Math.*, **1**, 157–162 (Vol. 12 of *Oeuvres de Poincaré*, pp. 130–133).
43. Pontrjagin L. and Schnirelmann L. (1932) Sur une propriété métrique de la dimension. *Ann. of Math.*, **33**, 156–162.
44. Quinqueton J. and Berthod M. (1981) A locally adaptive Peano scanning algorithm, *IEEE Trans.*, **PAMI-3** (4), 403–412.
45. Ramal R. (1984) Random walk statistics on fractal structures, *J. Statist Phys*, **36**(5/6), 547–560.
46. Read A. H. (1951–2) The solution of a functional equation, *Proc. Roy. Soc., Edinburgh Sec.*, **A63**, 336–345.
47. Scheidegger A. E. (1970) *Theoretical Geomorphology*, 2nd ed., Springer-Verlag, New York.
48. Stapleton H. J., Allen, J. P., Flynn C. P., Stinson D. G. and Kurtz S. R. (1980) Fractal form of proteins, *Phys. Rev. Lett.*, **45**(17), 1456–1459.
49. Tricot C. (1981) Douze définitions de la densité logarithmique, *C.R. Acad. Sci., Paris*, **293**, Série A, 549–552.
50. Tricot Jr. C. (1982) Two definitions of fractional dimension, *Math Proc. Camb. Philos. Soc.*, **91**, 57–74.
51. von Koch H. (1904) Sur une courbe continue sans tangente obtenue par une construction géométrique élémentaire, *Ark. Mat., Astronomie och Fysik*, **1**, 681–704.
52. Weierstrass F. (1872) Uber continuirliche functionen eines reellen arguments die für keinen werth des letzteren einen bestimmter differential quotienten besitzen, *Mathematische Werke*, **II**, 71–74.

Chapter 4 Dyadic interpolation

Gilles Deslauriers and Serge Dubuc
Department of Mathematics, University of Montreal

We describe a new interpolation method. The method depends on four parameters a, b, c and d and the starting point is a function $y(n)$ defined on the relative integers which we wish, if possible, to extend to the entire real line. If D_n denotes the set of dyadic rationals $m/2^n$ where m is a relative integer, $y(t)$ is already defined on D_0 and y is extended by induction to $D_1, D_2, D_3, \ldots$ as follows: if y has already been defined on D_n, $h = 2^{-n-1}$ and t belongs to D_{n+1} but not to D_n, the numbers $t - 3h, t - h$ and $t + 3h$ all belong to D_n and we put

$$y(t) = ay(t-3h) + by(t-h) + cy(t+h) + dy(t+3h) \qquad (1)$$

The extension $y(t)$ is thus defined on the set D of dyadic rationals p/q where p is a relative integer and q is an integer power of 2. In equation (1), the second author has studied the process of dyadic interpolation for the following choice of the parameters: $a = \mathrm{d} = -\frac{1}{16}, b = c = \frac{9}{16}$. The main question addressed in this work is that of determining conditions on the parameters a, b, c and d for the interpolation constructed on the dyadic numbers to be extendable by continuity to the whole real line.

4.1 Elementary properties of dyadic interpolation

We here give several elementary properties of the dyadic interpolation scheme (1), which can be checked in the same way as for the case $a = d = -\frac{1}{16}, b = c = \frac{9}{16}$ studied in [1].

Theorem 1 If $y(t)$ is the interpolation of the sequence $y(n)$ defined on D_0 and h is a negative integer power of 2, then the dyadic interpolation of the sequence $y(nh)$ is $y(th)$.

We now introduce the fundamental interpolant. We start with the sequence $F(n)$ defined on D_0 as follows: $F(0) = 1$ and $F(n)$ is zero for all other integer values of n. The dyadic interpolation $F(t)$ of this sequence will be called the fundamental interpolant and has the following properties:

Lemma 1 The fundamental interpolant vanishes outside the interval $(-3, 3)$.

Theorem 2 If $y(n)$ is a function defined on D_0, the dyadic extension of this sequence, $y(t)$, satisfies the identity $y(t) = \sum_{n=[t]-2}^{[t]+3} y(n)F(t-n)$, where $[t]$ denotes the integer part of the number t.

From the above identity and Theorem 1, one can obtain the following functional equation for the fundamental interpolant:

$$F(t/2) = F(t) + aF(t-3) + bF(t-1) + cF(t+1) + dF(t+3)$$

In Figures 4.1 to 4.4 we have plotted the graph of the fundamental interpolant for various choices of the parameters a, b, c and d.

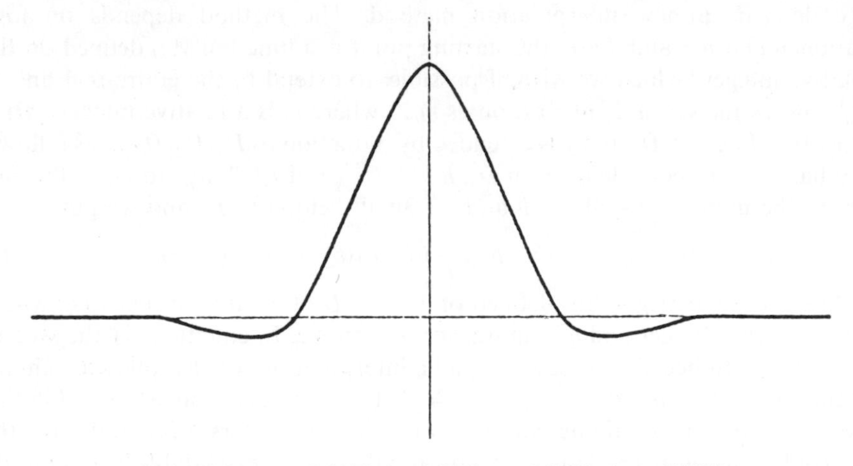

Figure 4.1 $a = -0.0625$, $b = 0.5625$, $c = 0.5625$, $d = -0.0625$

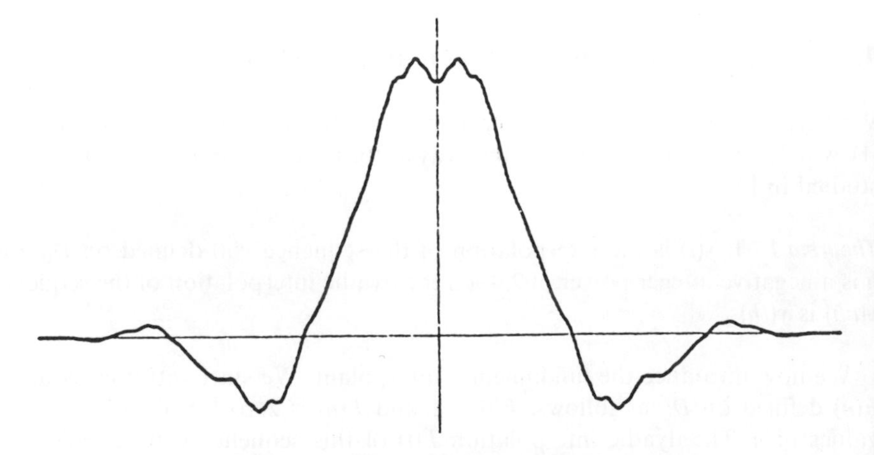

Figure 4.2 $a = -0.18$, $b = 0.68$, $c = 0.68$, $d = -0.18$

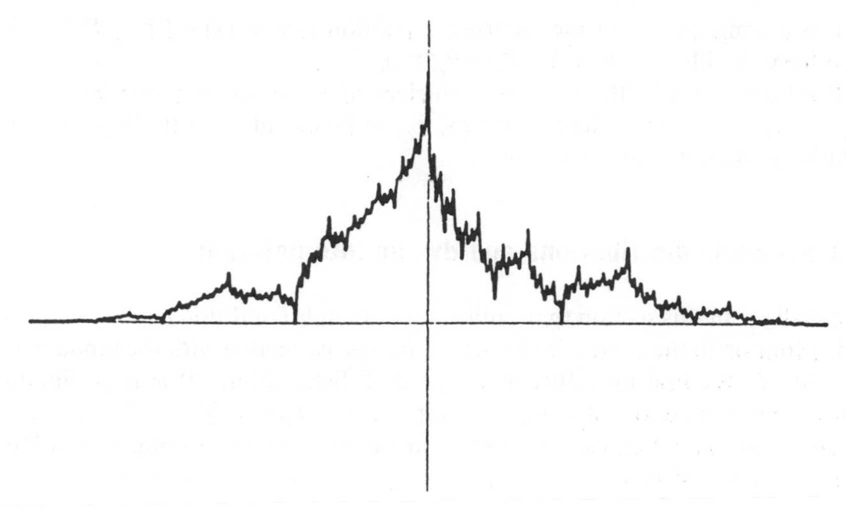

Figure 4.3 $a=0.3$, $b=0.1$, $c=0.4$, $d=0.2$

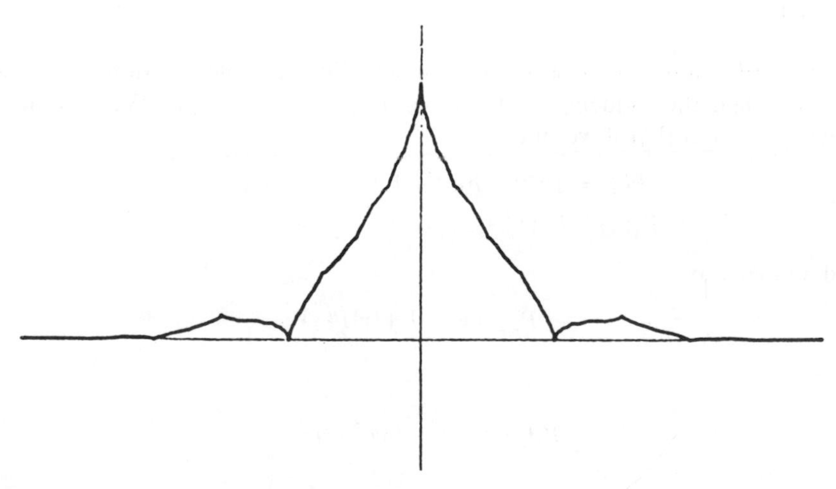

Figure 4.3 $a=0.1$, $b=0.4$, $c=0.4$, $d=0.1$

4.2 Rapid calculation of the fundamental function

We associate with the fundamental function a sequence of trigonometric polynomials defined in the following way: if $h=2^{-n}$, $P_n(x)=\sum_{k=-\infty}^{\infty} F(kh)e^{ikx}$.

We put $P(x)=a\,e^{3ix}+b\,e^{ix}+1+c\,e^{-ix}+d\,e^{-3ix}$ and $P_1(x)=P(x)$.

The recurrence relation $P_{n+1}(x)=P_n(2x)P(x)$ is established. The degree of P_n is $d_n=3(2^n-1)$.

It is a consequence of the recurrence relation that $P_n(x) = \prod_{k=0}^{n-1} P(2^k x)$. We also have the identity $P_{2n}(x) = P_n(x)P_n(2^n x)$.

If $P_n(x) = \sum c_k \mathrm{e}^{ikx}$, the Fourier coefficient of e^{ikx} in the polynomial $P_{2n}(x)$ is $\sum_{l=-\infty}^{\infty} c_{k+l2} c_{-l}$. This identity allows P_{2n} to be calculated rapidly and with a relatively small memory requirement.

4.3 Schwartz distributions and dyadic interpolation

We shall see in this section that, subject to the single condition $a + b + c + d = 1$, a distribution in the sense of Schwartz is always associated with the fundamental function F. We first introduce a sequence of distributions. If φ is an infinitely differentiable function of compact support, put $T_n(\varphi) = \sum_{x \in D_n} \mathrm{F}(x)\varphi(x)/2^n$. T_n is a distribution in Schwartz's sense, in fact a finite linear combination of Dirac masses supported in $D_n \cap [-3, 3]$.

Theorem 3 The sequence of distributions T_n converges to a distribution T of compact support. The Fourier transform $G(y) = T(\mathrm{e}^{-ixy})$ of the limit distribution satisfies the functional equation $G(2y) = G(y)P(y)/2$; furthermore, $G(0) = 1$.

Proof We show that, if $\varphi(x)$ is an infinitely differentiable function of compact support, then the sequence of numbers $T_n(\varphi)$ is convergent. We saw in the previous section that if we put

$$P(y) = a\,\mathrm{e}^{3iy} + b\,\mathrm{e}^{iy} + 1 + c\,\mathrm{e}^{-iy} + d\,\mathrm{e}^{-3iy}$$
$$P_1(y) = P(y)$$

and, inductively

$$P_{n+1}(y) = P_n(2y)P(y)$$

then

$$P_n(y) = \sum_{k=-\infty}^{\infty} F(k\,2^{-n})\mathrm{e}^{iky}$$

so that $F(k\,2^{-n})$ is the kth Fourier coefficient of $P_n(y)$. Whence

$$T_n(\varphi) = \frac{1}{2\pi} \int_{-\pi}^{\pi} \sum_k (k/2^n) P_n(y) \mathrm{e}^{-iky} 2^{-n}\,\mathrm{d}y \qquad (2)$$

The Poisson formula [3] states that if f and f'' are integrable over the real line, and if $g(y)$ is the Fourier transform of f, then $\sum_{k=-\infty}^{\infty} f(kh) = \sum_{k=-\infty}^{\infty} g(2\pi k/h)/h$. If we apply the Poisson formula with step h equal to 2^{-n} to the function $f(x) = \varphi(x)\mathrm{e}^{-ixz/h}$ and if $\psi(y)$ is the Fourier transform of $\varphi(x)$, then $\sum_k (k/2^n)\mathrm{e}^{-ikz} = 2^n \sum_k \psi((2\pi k + z)2^n)$. If we substitute from this identity into the integrand of equation (2), then

$$T_n(\varphi)=\frac{1}{2\pi}\int_{-\pi}^{\pi}\psi((2\pi k+y)2^n)P_n(y)\,\mathrm{d}y=\frac{1}{2\pi}\int_{-\infty}^{\infty}\psi(y\,2^n)P_n(y)\,\mathrm{d}y$$

whcnce

$$T_n(\varphi)=\frac{1}{2\pi}\int_{-\infty}^{\infty}\psi(y)P_n(y/2^n)\,\mathrm{d}y \tag{3}$$

This form of $T_n(\varphi)$ is the easiest for the study of the convergence of the sequence. Let $G_n(y)$ denote the function $P_n(y/2^n)/2^n=\prod_{k=1}^{n} P(y/2^k)/2$. From $P(0)=1+a+b+c+d=2$, it follows that, for every y in the interval $[-\pi,\pi]$, the inequality $|P(y)/2-1|\leqslant B|y|$ holds, provided B is sufficiently large. This inequality ensures the uniform convergence over every compact interval of the real line of the infinite product $\prod_{k=1}^{\infty} P(y/2^k)/2$. If $G(y)=\prod_{k=1}^{\infty} P(y/2^k)/2$, then $G_n(y)$ converges to $G(y)$ for every y.

The sequence of integrands in equation (3) thus converges pointwise to $\psi(y)G(y)$.

We now check that each of the integrands is dominated by one integrable function. We can find a constant M such that, for each y in $[-\pi,\pi]$, $|G_n(y)|\leqslant M$. We can use the number C, defined to be the maximum value of $|P(y)/2|$, to find a majorant of $G_n(y)$ valid for every real value of y. Indeed, if k is a natural number and if $|y|$ lies between the two values $\pi 2^k$ and $2\pi 2^k$, then $|G_n(y)|\leqslant MC^k$. This inequality enables us to find a number E such that $|G_n(y)|<M(1+|y|)^E$. Moreover, there exists a constant N such that $|\psi(y)|\leqslant N(1+|y|)^{-E-2}$. Each integrand is thus dominated by the integrable function $MN/(1+|y|)^2$.

By Lebesgue's dominated convergence theorem, the sequence $T_n(\varphi)$ converges to $\int\psi(y)G(y)\,\mathrm{d}y$ as n tends to infinity. According to Schwartz ([3], p. 74, theorem XIII), if a sequence of distributions converges pointwise, the limit functional is itself a distribution, and so the functional $T(\varphi)=\lim_{n\to\infty} T_n(\varphi)$ is a distribution.

Equally, the formula $T(\varphi)=\int\psi(y)G(y)\,\mathrm{d}y$ tells us that $G(y)$ is the Fourier transform of the distribution T. We observe the following about the function G: it is clear that the value of $G(0)$ is 1 and the expression for G allows us to say that $G(y)=G(y/2)P(y/2)/2$; whence

$$G(2y)=G(y)\,(a\,\mathrm{e}^{\mathrm{i}3y}+b\,\mathrm{e}^{\mathrm{i}y}+1+c\,\mathrm{e}^{-\mathrm{i}y}+d\mathrm{e}^{-\mathrm{i}3y})/2$$

Remarks The support of each distribution is contained in $[-3,3]$ and hence so also is the support of the limit distribution T.

The function F has been defined on the dyadics D. In the hope of extending F by continuity, when this is possible, we introduce, for each non-negative integer n, a piecewise linear function F_n defined as follows: if x is a dyadic number of the form $k/2^n$, $F_n(x)=F(x)$ and on the interval $[x,x+2^{-n}]$, F_n is linear. The next result shows that the sequence F_n is weakly convergent.

Theorem 4 If φ is an infinitely differentiable function of compact support and T is the limit distribution associated with F, then the sequence $\int F_n(x)\varphi(x)\,dx$ converges to $T(\varphi)$.

Proof Let $\tau(x)$ denote the 'hat' function: $\tau(x) = \max(1 - |x|, 0.$ $F_n(x) = \sum_k F(k/2^n)\tau(2^n(x - k/2^n))$. If μ_n denotes the combination $\sum_k 2^{-n}F(k/2^n)\delta_{k/2^n}$ of Dirac masses and if $\tau_n(x)$ is the function $2^n\tau(2^n x)$, F_n is the convolution of μ_n with τ_n.

If $\varphi(x)$ is an infinitely differentiable function of compact support and $\psi(y)$ is its Fourier transform, then $\int F_n(x)\varphi(x)\,dx = \int \psi(y)G_n(y)\,[\sin(y/2^{n+1})/(y/2^{n+1})]^2\,dy$. The term in square brackets is always dominated by 1 and tends to 1 pointwise as $n \to \infty$. The theorem of dominated convergence used in the preceding theorem gives us that $\int F_n(x)\varphi(x)\,dx$ tends to $\int \psi(y)G(y)\,dy = T(\varphi)$ as n tends to infinity.

Theorem 5 If $a = d$ and $b = c$ and $a \,\varepsilon\, (-\frac{3}{16}, \frac{1}{16})$, then the Fourier transform of the limit distribution T is integrable.

Proof The Fourier transform of T is

$$G(y) = \prod_{k=1}^{\infty} P(y/2^k)/2$$

Putting $y = 2^N z$ with $1 < z < 2$, we have

$$G(y) = \prod_{k=1}^{\infty} P(z\,2^N/2^k)/2 = \prod_{k=1}^{N} P(z\,2^{N-k})/2 \prod_{k=0}^{\infty} P(z/2^k)/2$$

and

$$|G(y)| \leqslant M \prod_{k=0}^{N-1} |P(z\,2^k)|/2 \text{ as } G \text{ is bounded on } [-\pi, \pi]$$

Let B_a denote the y-maximum of $|(1 - 8a\cos y + 8a\cos^2 y)|$. In view of the hypothesis that $a\varepsilon(-\frac{3}{16}, \frac{1}{16})$, it can be checked that $B_a < 2$.

From $P(z\,2^k) = (1 - 8a\cos 2^k z + 8a\cos^2 2^k z)(1 + \cos 2^k z)$ and

$$\prod_{k=0}^{N-1} \frac{(1 + \cos 2^k z)}{2} = \left[\frac{\sin(2^N z)}{2^N \sin(z/2)}\right]^2$$

it follows that

$$|G(y)| \leqslant M B_a^N \left[\frac{\sin(2^N z)}{2^N \sin(z/2)}\right]^2$$

If $B_a = 2^\beta$ (in this case, $0 < \beta < 1$),

$$|G(y)| \leqslant M' \frac{2^{\beta N}}{2^{2N}} \leqslant \frac{M''}{y^{2-\beta}}$$

for suitably chosen values of M' and M'' which are independent of y. It follows from this inequality that $G(y)$ is integrable.

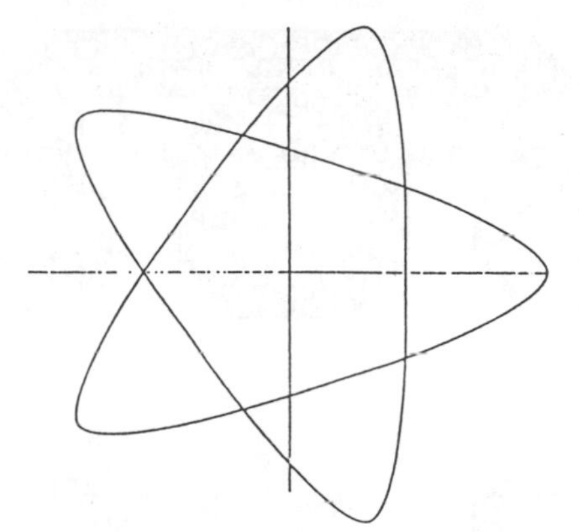

Figure 4.5

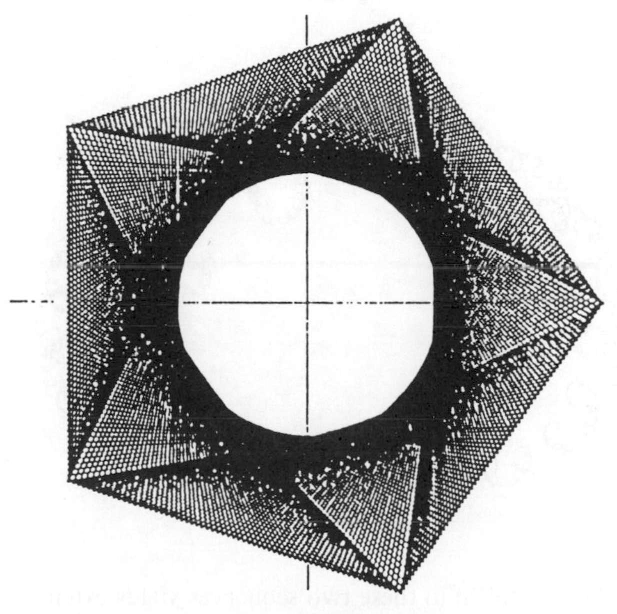

Figure 4.6

Corollary 6 If $a=d$ and $b=c$ and $a\varepsilon(-\frac{3}{16},\frac{1}{16})$, then the fundamental function $F(t)$ admits a continuous extension.

4.4 Applications to the interpolation scheme

Let $Z(n)$ be a sequence of complex numbers. Taking the real and imaginary parts separately, we obtain two real sequences $X(n)$ and $Y(n)$. The dyadic

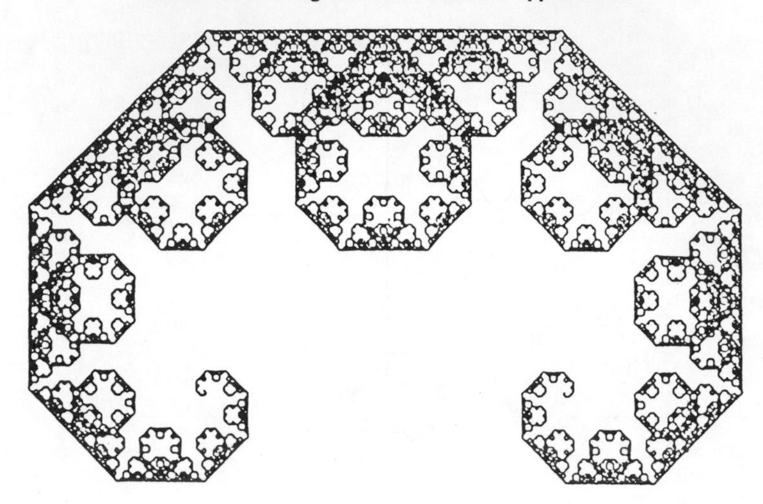

Figure 4.7

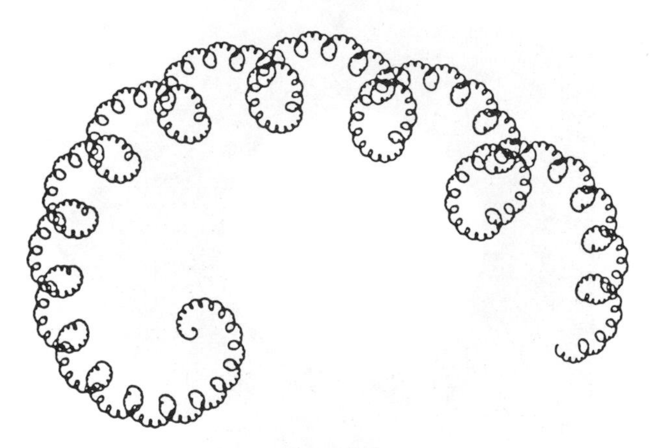

Figure 4.8

interpolation scheme applied to these two sequences yields extensions $X(t)$ and $Y(t)$ respectively. $Z(t)$ is thus the complex function $X(t) + \mathrm{i}\,Y(t)$. We give here the graphical representation of just one sequence of complex numbers for two choices of the parameters a, b, c and d.

The sequence employed will be $Z(n) = \mathrm{e}^{\mathrm{i}4\pi n/5}$, i.e. that generated by the second fifth root of unity. Figures 4.5 and 4.6 represent the cases $a = d = 0.0625$, $b = c = 0.5625$ and $a = d = 0.5$, $b = c = 0$ respectively.

Finally, if $Y(n) = 0$ for $n = 0$ and $Y(0) = 1$, we give two graphical representations in the complex plane. The first, Figure 4.7, is Gosper's curve in the complex plane **C**. It is obtained using the interpolation scheme when $a = d = 0$, $b = \frac{1}{2} + \mathrm{i}/2$ and $c = \frac{1}{2} - \mathrm{i}/2$, and t varies in [0, 1].

The last curve, the 'lace pattern' (Figure 4.8), is obtained by taking $a = d = 0$, $b = \frac{1}{5} + i/3$ and $c = \frac{4}{5} - i/3$, with t again varying in the interval $[0, 1]$.

The last two curves belong to the class of von Koch–Mandelbrot curves [2].

References

1. Dubuc, S. (to appear) Interpolation through an iterative scheme, *J. Math. Anal. Appl.*
2. Mandelbrott, B. (1984) *Les Objects Fractals*, 2nd ed. Flammarion.
3. Schwartz, L. (1966) *Théorie des Distributions*, Hermann, Paris.

Chapter 5 Stochastic processes and covering procedures

M. Weber
Department of Mathematics, University of Strasbourg

5.1 Some covering procedures—their interrelationships and examples

In this chapter we present some of the most important covering procedures and the relationships between them. We illustrate them with the help of examples whose source is, most often, the study of stochastic processes or harmonic analysis. The following section deals with the notion of Hausdorff dimension and, to start with, we have followed J.-P. Kahane's excellent treatment of it [21].

5.1.1 Hausdorff measure and dimension

The following notion was introduced by Hausdorff in 1919. Consider a separable metric space E with metric d and let $\mathscr{H}$ be the set of increasing functions from $\mathbf{R}^+$ onto itself. Note that such functions are necessarily continuous and vanish at 0. With each $\varepsilon > 0$, one associates the set of coverings of E by balls B_n of diameter less than or equal to ε; thus

$$E \subset \bigcup_n B_n, \qquad \text{diam } B_n \leqslant \varepsilon$$

We write $H(\varepsilon) = \inf \sum h(\text{diam } B_n)$, the infimum being taken over the set of all coverings of the above type. Clearly $H(\varepsilon)$ increases as $\varepsilon \downarrow 0$, and $\lim_{\varepsilon \downarrow 0} H(\varepsilon) = H$ satisfies $0 < H < \infty$ and is called the Hausdorff measure of E with respect to the defining function h. We put

$$H = m_h(E)$$

This is an outer measure (in this connection, see [32], pp. 40–67, for example). When $h_\alpha(t) = t^\alpha$, the mapping $\alpha \to m_{h_\alpha}(E)$ is decreasing and is everywhere equal to either 0 or ∞ with the possible exception of just one point. In this case, we define

$$\alpha_0 = \inf\{\alpha\colon m_{h_\alpha}(E) = 0\} = \sup\{\alpha\colon m_{h_\alpha}(E) = \infty\}$$

The real number α_0 defines the Hausdorff dimension of E:

$$\alpha_0 = \dim_H(E)$$

Some relationships that are valid in any euclidean space are

$$\forall \lambda > 0, \qquad m_{h_\alpha}(\lambda E) = \lambda^\alpha m_{h_\alpha}(E)$$
$$m_{h_\alpha}(E + x) = m_{h_\alpha}(E)$$

The following theorem provides a first characterization of this notion.

Theorem 1 (O. Frostman, [12], *p.* 89) Let E be a compact subset of $\mathbf{R}^d$ and let h be a specifying function satisfying the condition

$$(\Delta_2) \qquad h(2t) = O(h(t)), \qquad \text{as } t \to 0$$

In these circumstances, we have the equivalence

$$(m_h(E) > 0) \Leftrightarrow \begin{cases} \text{there exists a positive non-zero measure } \mu \text{ on} \\ E \text{ such that } \mu(B) \leqslant h(\text{diam } B) \text{ for every} \\ \text{ball } B \end{cases}$$

Remark The implication ($\Leftarrow$) is true in general. Indeed, let $(B_n)_n$ be a covering of E. Then

$$0 < \mu(E) < \sum_n \mu(B_n) \leqslant \sum_n h(\text{diam } B_n)$$

Here is a second characterization:

Theorem 2 In every metric space, we have the following equivalence:

$$(m_h(E) = 0) \Leftrightarrow \begin{cases} \text{there exists a sequence of balls } (B_n)_n \text{ such that} \\ \sum_n h(\text{diam } B_n) < \infty \text{ and } E \subset \lim_{n\to\infty} \sup B_n \end{cases}$$

Proof Suppose that $m_h(E) = 0$. By definition, for each natural number n, there exists a sequence of balls $(B_i^{(n)})_i$ such that

$$E \subset \sum_{i=1}^{\infty} B_i^{(n)} \qquad \text{and} \qquad \sum_{i=1}^{\infty} h(\text{diam } B_i^{(n)}) \leqslant 2^{-n}$$

Reindexing the double sequence $(B_i^{(n)})_{i,n}$ as the single sequence $(E_p)_p$, we have

$$E \subset \lim_{n\to\infty} \sup E_n \qquad \text{and} \qquad \sum_{n=1}^{\infty} h(\text{diam } E_n) < \infty$$

Conversely, let $(B_n)_n$ be such a sequence of balls and let $\varepsilon, \delta > 0$. Since diam $B_i \leqslant \delta$ for all but a finite number of indices, there exists an integer $N = N(\varepsilon, \delta)$ such that

$$\forall i \geqslant N, \qquad \text{diam } B_i \leqslant \delta \qquad \text{and} \qquad \sum_{i=N}^{\infty} h(\text{diam } B_i) \leqslant \varepsilon$$

It follows from this, as $E \subset \bigcup_{i=N}^{\infty} B_i$, that $m_h(E) \leqslant \varepsilon$, whence the result, on letting ε tend to zero.

It can be verified without difficulty that, in the real line, the Cantor ternary set has Hausdorff dimension equal to $\log 2/\log 3$.

In 1919, Hausdorff gave a simple method for constructing on the real line a Cantor set E such that

$$m_h(E) = 1$$

as long as h is concave (in fact, as long as h satisfies $h(2u) \leqslant 2h(u)$). This method is illustrated by Figure 5.1.

In fact, A. Dvoretsky ([5], p. 15) has since shown that there exists a compact set $S \subset \mathbf{R}^n$ such that

$$0 < m_h(S) < \infty$$

if and only if

$$\liminf_{x \to 0} \frac{h(x)}{x^n} > 0$$

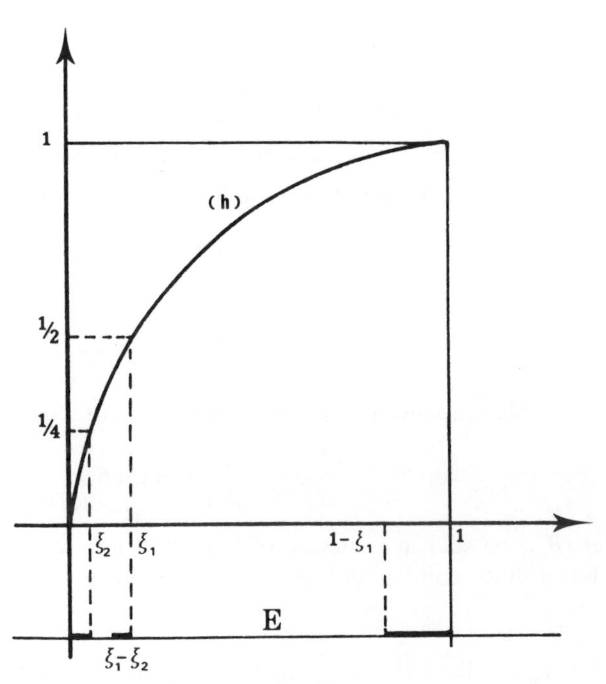

Figure 5.1

5.1.2 Some applications

Example 1 Let $B=(B(t), 0<t<\infty)$ be a linear brownian motion. Let $B_1, \ldots, B_n$ be n independent copies of B and put

$$W_n=(B_1, B_2, \ldots, B_n)$$

the multidimensional brownian motion thereby obtained. It is known that the trajectories of this process are, almost surely, continuous but not differentiable. Let E denote the image of the interval $[0,1]$ under W_2. It can be shown that E is almost surely a compact set of Lebesgue measure zero but that

$$\dim_H(E)=2 \text{ almost surely}$$

In fact, ([4], p. 434), if $h(t)=t^2 \log\log 1/t$, we have the somewhat sharper result

$$0<m_h(E)<\infty \text{ almost surely}$$

Similarly, the notion of Hausdorff dimension may be usefully applied in the study of the level sets of a process; i.e. in the present context, the sets, defined for $x \varepsilon \mathbf{R}^n$,

$$E_x=\{t \in \mathbf{R}: W_n(t)=x\}$$

and especially matters concerning the eventual multiple points of the graph of W_n. In this connection, recall the classical results of A. Dvoretsky, P. Erdös and S. Kakutani.

Theorem 3 ([6–8])

(a) The graph of W_2 possesses, almost surely, points of every order of multiplicity.
(b) The graph of W_3 possesses, almost surely, double points but no triple points.
(c) For every natural number $n>4$, the graph of W_n almost surely has no double points.

We set

$$\forall k \geqslant 1, M_k^n=\{(t_1, \ldots, t_k) \in \mathbf{R}^k, t_i \neq t_j, W_n t_1=\cdots=W_n(t_k)\}$$

We then have:

Theorem 4 ([29]) Almost surely, $\dim_H(M_k^n)=\max([k-(k-1)/2]n, 0)$

Remarks These theorems have only recently been extended to the case of fractional brownian fields using totally different techniques. Consider a d-dimensional fractional brownian field $X_{N,\beta}$ indexed over $\mathbf{R}^N$, of order $0<\beta<1$. Its coordinates are independent and obey the same law as the usual fractional brownian motion of order β introduced by B. Mandelbrot and J. Van Ness [27]. A. Goldman showed in [13] that for every natural number $k \geqslant 1$, if the parameters N, β, k and d satisfy the relationship $Nk<\beta(k-1)d$, then, almost surely, $X_{N,\beta}^d$ has no points of multiplicity k, whilst if $Nk>\beta(k-1)d$, the trajectories of $X^d_{N,\beta}$ almost surely do possess points of multiplicity k. The question of what happens when $Nk=(k-1)d$ is still unsettled, except for the already treated case of ordinary brownian motion on the line.

We now put

$$M_k = M_k(d, N, \beta) = \{(t_1, \dots, t_k) \varepsilon \mathbf{R}^{Nk}, \text{ all } t_i \text{ distinct and } X^d_{N,\beta}(t_1) = \dots = X^d_{N,\beta}(t_k)\}$$

We can then prove:

Theorem 5 ([44]) When $Nk > (k-1)d$, almost surely, $\dim_H(M_k) = Nk - (k-1)\beta d$.

Example 2 (Simultaneous diophantine approximation) Let $\mathbf{I}$ denote the set of irrational numbers and let the map $\omega: \mathbf{R}^+ \to \mathbf{R}^+$ be decreasing and vanishing at infinity. For a given natural number r, consider the r-tuples $(\theta_1, \dots, \theta_r) \in \mathbf{I}^r$ such that the inequalities

$$(E_r) \qquad \forall i = 1, 2, \dots, r, \left|\theta_i - \frac{p_i}{q}\right| < \omega(q), \qquad p_i, q \in \mathbf{Z},$$

hold for infinitely many integers q. For $r=1, \omega(x) = 1/x$, we recover the classical property of approximation of irrationals by irreducible fractions p/q to within an error strictly less than $1/q$.

In the present context, we have the following difficult theorem of Jarnik, proved in 1931 under fairly weak hypotheses on ω and h.

Theorem 6 ([18] and [19]) Let $A_r(\omega) \subset \mathbf{I}^r$ be the set of solutions of (E_r) and let h be a specifying function. In these circumstances, we have the following criterion:

$$\{m_h(A_r(\omega)) = 0, \text{ (resp.} = \infty)\} \Rightarrow \left\{\int_1^\infty h(2\omega(x))x^n \, dx < \infty, \text{ (resp.} = \infty)\right\}$$

Example 3 (Section theorems) These results are due to J. M. Marstrand [30] and are concerned with the sections of a plane measurable set. Following his terminology, we shall say that a plane Borel set E is an s-set if $0 < m_{h_s}(E) < \infty$. In this case, if $s > 1$ (clearly $s \leqslant 2$), the Lebesgue measure of the projection of E onto the line D_θ emanating from the origin is positive and this is true for almost every θ (Figure 5.2).

Now let x be a point of E and consider the intersection of E with the line $D_{\theta,x}$ of slope θ passing through x (Figure 5.3).

We have the significantly stronger result:

Theorem 7 ([30], p. 275) For an s-set E with $s > 1$, for almost every x in E and almost every real θ,

$$\dim_H(E \cap D_{\theta,x}) = s^{-1} \quad \text{and} \quad m_{h_{s-1}}(E \cap D_{\theta,x}) < \infty$$

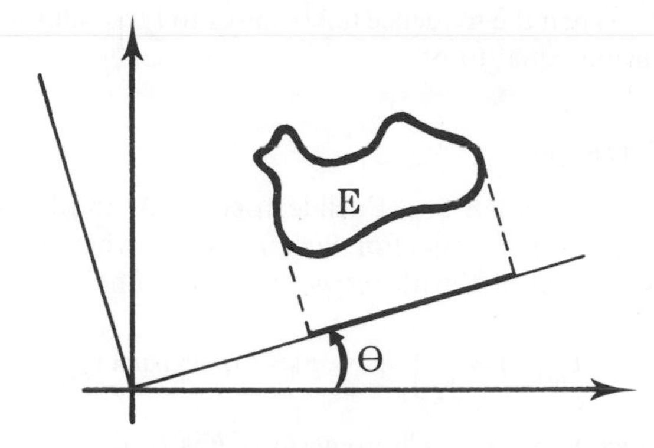

Figure 5.2

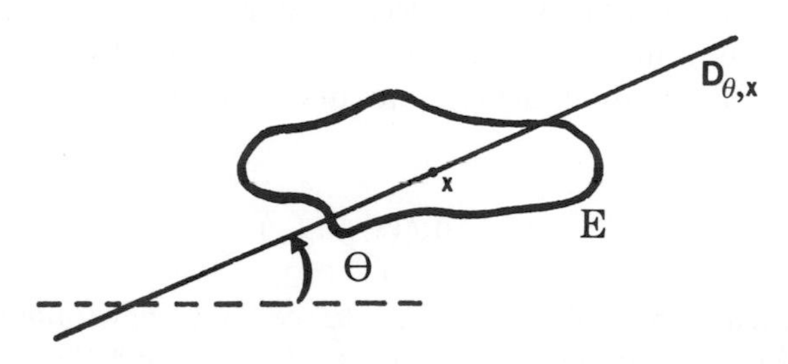

Figure 5.3

Example 4 (Rademacher series) Consider a sequence $(R_n)_{n>1}$ of independent Rademacher random variables. To fix ideas, one can choose, for example, the sequence $(R_n(x))_{n\geqslant 1}$ of coefficients of the binary expansion of $x \in [0, 1]$ with Lebesgue space as test space. Recall that the binary expansion of x is unique if one excludes the following kind of possibility:

$$R_{n-1}(x) = 0 \qquad \text{and} \qquad 1 = R_n(x) = R_{n+1}(x) = \cdots$$

which is, in any case, just a way of writing something that can be avoided by putting, instead, $R_{n-1}(x) = 1$. Let, then, $(a_i)_{i\geqslant 1}$ be an element of $c_0 \backslash l_1$, so that $|a_i| \to 0$ and $\sum_{i\geqslant 1} |a_i| = \infty$. It is an easy consequence of the theorem of Kaczmarz and Steinhaus ([19], theorem 2 with $a = b$) that the level sets of the Rademacher series

$$S(x) = \sum_{i=1}^{\infty} a_i R_i(x)$$

i.e. the sets $S^{-1}(c)$, $c \in \mathbf{R}$, all have the power of the continuum. This result can be sharpened, viz. in [2], theorem 1, W. A. Beyer proves:

Theorem 8 ([2]) When the sequence (a_i) belongs to $l_2 \setminus l_1$, all the level sets have Hausdorff dimension equal to one.

5.1.3 Capacity dimension

Consider a compact subset E of a Euclidean space X and let m be a positive Borel measure giving E mass one. For each $0 < \alpha < n$, where n is the dimension of X, the energy integral of E with respect to m is defined by

$$I_{\alpha,m}(E) = \int_E \int_E \|x - y\|^{\alpha - n}\, \mathrm{d}m(x)\, \mathrm{d}m(y)$$

where $\|\cdot\|$ is the euclidean norm. The α-energy of E is, by definition, the quantity

$$W_\alpha(E) = \inf I_{\alpha,m}(E)$$

where the infimum is taken over all distributions of unit mass over E. This notion has its origin in electrostatics; unit electric charge on an isolated conductor $E \subset \mathbf{R}^3$ spreads itself in such a way that the associated distribution μ_e, called the equilibrium distribution, achieves its minimum $W_2(E)$, i.e. it satisfies the inequality

$$\iint \|x - y\|^{-1}\, \mathrm{d}\mu_e(x)\, \mathrm{d}\mu_e(y) = W_2(E)$$

Regarding the name 'energy', it is not hard to see (in the newtonian case) that the quantity $W_2(E)$ corresponds to the work needed to bring the charge, Q, from infinity to the compact set E, i.e. work performed in the course of movement through the force-field associated with the potential $G_2(\mu_e)(x) = \int_{\mathbf{R}^3} \|x - y\|^{-1}\, \mathrm{d}\mu_e(y)$, called the equilibrium potential.

Recall that the equilibrium measure is unique; in other words, there exists a unique probability measure μ_e satisfying

$$I_{\alpha,\mu_e}(E) = W_\alpha(E)$$

and that the equilibrium potential is equal to $W_\alpha(E)$, μ_e-almost everywhere on E.

The α-capacity of E is the quantity

$$C_\alpha(E) = \begin{cases} [W_\alpha(E)]^{-1/\alpha} & \text{if } W(E) < \infty \\ 0 & \text{otherwise} \end{cases}$$

The mapping $\alpha \to C_\alpha(E)$ is decreasing, and so the capacity dimension of E is unambiguously defined by

$$\dim_C(E) = \begin{cases} 0 & \text{if } C_\alpha(E) = 0 \quad \forall \alpha > 0 \\ s > 0 & \text{if } C_\alpha(E) = 0 \quad \forall \alpha > s \\ & \text{and } C_\alpha(E) > 0 \quad \forall 0 < \alpha < s \end{cases}$$

This is linked to the Hausdorff dimension via:

Theorem 9 (O. Frostman [12]) For every compact subset E of a euclidean space, we have

$$\dim_C(E) = \dim_H(E)$$

It goes without saying that this notion of dimension generalizes without difficulty to any separable metric space, as S. J. Taylor has pointed out in [40]. Let F be a Borel subset of such a space, let $M_1^+(F)$ be the set of positive Borel measures of mass one on F and, lastly, let $J:\mathbf{R}^+ \to \mathbf{R}^+$ be an unbounded decreasing function.

The J-energy of F is the quantity

$$W^J(F) \quad \inf_{\mu \in \mathbf{M}_1^+(F)} \iint J(d(s,t))\,\mathrm{d}\mu(s)\,\mathrm{d}\mu(t)$$

where d is the metric on F. The generalized J-capacity of F is defined as follows:

$$C^J(F) = \begin{cases} J^{-1}(W^J(F)) & \text{if } W^J(F) < \infty \\ 0 & \text{otherwise} \end{cases}$$

The capacity is intrinsically linked with the chosen metric. Regarding this, consider the sequence

$$D_n^J = \inf_{x_1, x_2, \ldots, x_n \in F} \frac{1}{n(n-1)} \sum_{1 \le i < j \le n} J(d(x_i, x_j)).$$

This sequence is decreasing and its limit $D^J(F)$ determines what is customarily called the transfinite diameter of F of order J.

We then have a metric characterization of the capacity in the form of:

Theorem 10 (G Szegö and S. Polya [12]) For every compact set F, we have $D^J(F) = W^J(F)$.

Proof For each choice of $(x_1, \ldots, x_n) \in F^n$, the following inequality is satisfied:

$$\frac{1}{n(n-1)} \sum_{1 \le i < j \le n} J(d(x_i, x_j)) > D_n^J(F)$$

Let μ be a probability measure on F and let us integrate the above expression with respect to $\otimes_1^n \mu$. We obtain

$$\iint J(d(x,y))\,\mathrm{d}\mu(x)\,\mathrm{d}\mu(y) \geqslant D_n^J(F) \geqslant D^J(F)$$

and, finally, $W^J(F) \geqslant D^J(F)$.

Conversely, the function

$$(x_1, \ldots, x_n) \to \frac{1}{n(n-1)} \sum_{1 \le i < j \le n} J(d(x_i, x_j))$$

is lower semi-continuous on the compact set F^n, and therefore attains its minimum at some point $(x_1^n, \ldots, x_n^n)$. Consider the sequence of measures

$$\mu_n = \frac{1}{n} \sum_{i=1}^{n} \delta_{x_i^n}$$

Since F is compact, the sequence μ_n is weakly compact. Taking a subsequence if necessary, we can assume that the sequence μ_n converges weakly to a probability measure μ on F. Since

$$D_n^J(F) = \iint J(d(x, y)) \, \mathrm{d}\mu_n(x) \, \mathrm{d}\mu_n(y)$$

we will therefore have, for arbitrary $\varepsilon > 0$,

$$\begin{aligned} D^J(F) &\geqslant \lim_{n \to \infty} \iint \inf(\varepsilon, J(d(x, y)) \, \mathrm{d}\mu_n(x) \, \mathrm{d}\mu_n(y) \\ &= \iint \inf(\varepsilon, J(d(x, y)) \, \mathrm{d}\mu(x) \, \mathrm{d}\mu(y) \end{aligned}$$

from which the result follows on letting ε tend to zero.

5.1.4 The dimension of Schnirelman–Pontjrajin and of Kolmogorov

Consider a compact metric space E and let $\varepsilon > 0$. By the Heine–Borel theorem, E can be covered by a finite number of balls of diameter less than or equal to ε. Let $N(\varepsilon)$ denote the smallest number of balls in such a covering. The metric order of E, introduced by Schnirelman and Pontjrajin [37], or, alternatively, the metric entropy of E is intended to compare $N(\varepsilon)$ with $\varepsilon^{-d}, d > 0$; whence the definition

$$d(E) = \limsup_{\varepsilon \to 0} \frac{\log N(\varepsilon)}{\log 1/\varepsilon}$$

Of course, compact metric spaces with finite metric entropy are somewhat rare. Thus, on the real line, if $E = [0, 1]$ is endowed with the distance $d(s, t) = \sigma(|s - t|)$, where σ is a concave increasing function vanishing at zero, σ would have to increase regularly in the sense of J. Karamata with strictly positive exponent, i.e $\sigma(t) \propto t^\alpha, t \to 0, \alpha > 0$, which excludes various interesting situations such as, for example,

$$\sigma(t) = \left(\log \frac{1}{t} \right)^{-b}, \qquad b > 0$$

or

$$\sigma(t) = \exp\left[-\left(\log \frac{1}{t} \right)^c \right], \qquad c > 0, \text{ for example.}$$

With the function $N(\varepsilon)$, one generally associates the following:

$$M(\varepsilon) = \begin{cases} \text{cardinal number of a maximal family of points of } E \text{ that are} \\ \text{`}\varepsilon\text{–}d\text{-distinguishable' or form an `}\varepsilon\text{-net'; i.e. the distance between} \\ \text{any two distinct points of the family is strictly greater than } \varepsilon \end{cases}$$

It is easily shown that $M(\varepsilon) \geqslant N(\varepsilon)$.

In a similar way, A. Kolmogorov ([26]) associates a second metric entropy.

Finally, let us draw attention to another covering function which seemed particularly interesting to us and whose origin probably goes back to G. Bouligand [3]. For $a, b > 0$, we put

$$N(a, b) = \sup \operatorname{card}(S \cap B(y, a))$$

the maximum being taken over all points y of E and all a-nets S of E.

5.1.5 The Menger–Urysohn dimension

This dimension is defined inductively as follows:

(a) The empty set has dimension -1.
(b) A topological space is of dimension n if n is the largest natural number such that each point of the space has arbitrarily small neighbourhoods with frontiers of dimension less than n.

Thus a space is of dimension 0 if every point p of the space has arbitrarily small neighbourhoods with an empty frontier, i.e. for every neighbourhood U of p, there exists a neighbourhood V of p such that

$$V \subset U$$
$$\operatorname{fr}(V) = \phi$$

The set of irrational numbers, the Cantor set and any countable space are of dimension 0. Every non-empty subset of a space of dimension 0 is itself zero dimensional.

It is a by no means trivial result that a euclidean space of dimension n in the usual sense is of dimension n in the sense of Menger–Urysohn. Most of the properties of this dimension are dealt with in the book *Dimension Theory* by W. Hurewicz and Wallman [16].

We mention two properties that by themselves justify the interest in this notion.

Theorem 11 ([16], theorem VI.7, p. 91] Let X and Y be two topological spaces and let $f: X \to Y$ be a closed mapping. Suppose further that

$$\dim X - \dim Y = k > 0$$

Then there exists a point of Y whose inverse image under f has dimension $\geqslant k$.

The following theorem establishes an important link with the Hausdorff dimension.

Theorem 12 [(16], theorem VII.5 and p. 107) Let X be a separable metric space. Then

$$\dim X = \inf \dim_H X'$$

where the minimum is taken over the set of spaces X' homeomorphic to X.

5.2 Applications to the study of the regularity of stochastic processes

5.2.1

Let $(\Omega, \mathscr{A}, P)$ be a probability space, let T be some set and $X = \{X(\omega, t), \omega \in \Omega, t \in T\}$, a stochastic process indexed by T—in other words, a family of real random variables (r.r.v.) indexed by T.

The study of the regularity of the process X is the study of the orbits of X, i.e. the mappings $t \to X(\omega, t), \omega \in \Omega$. Generally, one establishes properties common to almost all the orbits (almost sure properties), such as, for example, their being almost surely majorized, their continuity, their Lipschitz character, etc.

It is thus well known that classical brownian motion is almost surely continuous, but not differentiable. The way one proceeds varies very little. Under hypotheses on the increments $X_s - X_t$ of the process, concerning either their moments or the order of magnitude of $P\{|X_s - X_t| > u\delta(s, t)\}$, as $u \to \infty$, where δ is a certain distance on T, one states conditions on the geometric nature of (T, δ), which mean that (T, δ) is not too 'big' and then imply that the family of the process, determined by the hypotheses, has 'good' versions with orbits that are continuous or majorized, as the case may be. The theorem below provides a good illustration of this.

Theorem 13 (Kolmogorov; see [26], 35–3, pp. 513–519) Let $\{X_t, 0 \leqslant t \leqslant 1\}$ be a random process. We suppose that $0 < p < \infty$ and $\delta > 0$ are such that

$$\forall t, s \in [0, 1], \qquad \mathbf{E}\{|X_s - X_t|^p\} \leqslant |t - s|^{1+\delta}$$

Then the process $\{X_t, 0 \leqslant t \leqslant 1\}$ possesses a version with continuous orbits.

This result has recently been refined by M. G. Hahn, I. Ibraginov and N. Kôno.

Theorem 14

(a) Let φ: $[0, 1] \to \mathbf{R}^+$ be an increasing function vanishing at zero such that $\varphi(x)/x$ is decreasing and such that

$$(I_p) \qquad \int_0^1 \frac{\varphi(t)}{t^{1+1/p}} \mathrm{d}t < \infty$$

Under these conditions, we have

$$(K_p) \qquad \begin{cases} \text{Every process } X_t, 0 \leqslant t \leqslant 1 \text{ that satisfies, } \forall t, s \in [0,1], \\ (\mathbf{E}|X_s - X_t|^p)^{1/p} \leqslant \varphi(|t - s|) \text{ necessarily admits a version} \\ \text{with continuous orbits.} \end{cases}$$

(b) ([27] and [17]) Let p be such that $1 < p < \infty$. If φ satisfies (K_p), then it necessarily also satisfies the integral condition (I_p).

Note Part (b) of the statement says that if the integral condition (I_p) fails, then there exists an element X_0 of the family determined by the condition

$$\forall t, s \in [0,1], \qquad (\mathbf{E}|X_s - X_t|^p)^{1/p} \leqslant \varphi(|t - s|)$$

whose orbits are almost surely discontinuous.

This kind of statement is satisfying from a theoretical point of view since it characterizes a property, namely the continuity of the orbits for the family of a process; nevertheless, the process employed, X_0, is still far too artificial since one has to take Lebesgue space for the probability space and to choose a random Fourier series in order to construct X_0.

Theorem 14 does not suffer from this kind of handicap; on the other hand, it is markedly less precise.

For a general T, the conditions are expressed by means of entropy functions $N_d(T, \varepsilon)$ where d is a distance on T, usually determined by the process X. Thus, let φ be a Young function and suppose that the following hypothesis is satisfied:

$$(H_\varphi) \qquad \forall s, t \in T, \qquad \mathbf{E}\,\varphi\left(\left|\frac{X_s - X_t}{\mathrm{d}(s,t)}\right|\right) \leqslant 1$$

Consider the integral

$$I_\varphi = \int_0^{\mathrm{diam}(T,d)} \varphi^{-1}(N_d(T,u))\,\mathrm{d}u$$

In this setting we have:

Theorem 15 For the process X to possess a version with continuous orbits, it is sufficient that I_φ be convergent.

Note In the gaussian case (recall that a random process X is gaussian if and only if every finite linear combination of the r.r.v.'s X_t obeys a gaussian law), this condition reduces to

$$\int_0^{\mathrm{diam}(T,d)} \sqrt{\log N_d(T,u)}\,\mathrm{d}u < \infty$$

A remarkable theorem of X. Fernique ([9], theorem 8.1.1) shows that this condition characterizes the regularity of stationary gaussian processes indexed by $\mathbf{R}^n$.

Just as the preceding theorem concerns the so-called 'convex' case (because φ is convex), the next theorem deals with the 'non-convex' case i.e. the families of processes satisfying the natural hypothesis ($\mathscr{H}$) that there exists an r.r.v. Λ_0 such that for all s and t in T and every $x>0$

$$P\{|X_s - X_t| \geqslant x\,\delta(s,t)\} \leqslant P\{\Lambda_0 > x\}$$

where δ is some fixed distance on T.

We write $\Gamma(u)$ for the decreasing function on $[0, 1]$ which satisfies the condition $P\{\Lambda_0 > \Gamma(u)\} = u$. We then have $R_{\Lambda_0}(x) = R(x) = x\,\mathbf{E}\{\Lambda_0 1_{\Lambda_0 > \Gamma(1/x)}\}$ is decreasing on $\mathbf{R}_+{}^*$.

We prove the following theorem:

Theorem 16 ([42], p. 394) Let $\{X_t, t\in T\}$ be a δ-separable stochastic process satisfying the hypothesis ($\mathscr{H}$). Under these conditions, for X to be almost surely majorized, it is sufficient that the integral

$$\int_0^{\operatorname{diam}(T,\delta)} R[N_\delta(T,u)\,\mathrm{d}u \tag{1}$$

be convergent.

Then, at each point t_0 of T, the process X is almost surely continuous and for every $0<a<1$, we have

$$\mathbf{E}\left\{\sup_{t\in T}|X_t - X_{t_0}|\right\} < 4\int_0^{\operatorname{diam}(T,d)} R[N_\delta(T,u)]\,\mathrm{d}u \tag{2}$$

$$P\left\{\sup_{t\in T}|X_t - X_{t_0}|\right\} > 4\int_0^{\operatorname{diam}(T,d)} R\left[\frac{N_\delta(T,u)}{a}\right]\mathrm{d}u\right\} \leqslant a \tag{3}$$

Examples

(a) *The power case.* Let $p>1$ and let $X:(T,\delta)\to\mathbf{R}$ be a δ-separable random process satisfying one of the following two conditions:

$$\sup_{T\times T}\mathbf{E}\left|\frac{X_s - X_t}{\delta(s,t)}\right|^p < \infty \tag{4}$$

$$\limsup_{x\to\infty} x^p \sup_{T\times T} P\{|X_s - X_t| > x\delta(s,t)\} < \infty \tag{5}$$

In this context, condition (1) takes the form

$$\int_0^{\operatorname{diam}(T,\,\delta)} N_\delta(T,u)^{1/p}\,\mathrm{d}u < \infty \tag{6}$$

Furthermore, (3) yields

$$\forall \varepsilon > 0, \qquad \mathbf{E}\left\{\frac{\sup_{T\times T}|X_s - X_t|^p}{\log \sup_{T\times T}|X_s - X_t|^{1+\varepsilon}}\right\} < \infty \tag{7}$$

as well as,

$$\forall \varepsilon > 0 \text{ and } \forall x \text{ sufficiently large,}$$

$$P\left\{\sup_{\delta(s,t_0)<\varepsilon} |X_s - X_{t_0}|\right\} > x \int_0^\varepsilon N(B_\delta(t_0,\varepsilon),\omega)^{1/p}\,\mathrm{d}u\Bigg\} < \left(\frac{5}{2}\right)^p x^{-p} \tag{8}$$

In what follows, we have put

$$\forall x > 0, \forall R > 0, \forall \tau > 0, \Phi_{R,\tau}(x) = \exp(-Rx^\tau)$$

(b) *The exponential case.* Let $X:(T,\delta)\to\mathbf{R}$ be a δ-separable random process satisfying one of the following two conditions:

$$\sup_{T\times T} \mathbf{E}\left\{\Phi_{R,\tau}\left(\left|\frac{X_s - X_t}{\delta(s,t)}\right|\right)\right\} < \infty \tag{9}$$

$$\limsup_{x\to\infty} \sup_{T\times T} \frac{P\{|X_s - X_t| x\delta(s,t)\}}{\Phi_{R,\tau}(x)} < \infty \tag{10}$$

Condition (1) here becomes

$$\int_0^{\operatorname{diam}(T,\delta)} [\log \mathrm{N}(T,u)]^{1/\tau}\,\mathrm{d}u < \infty \tag{11}$$

In addition to the conclusion of Theorem 16, there exists $\alpha_0 > 0$ such that

$$\forall \alpha < \alpha_0, \qquad \mathbf{E}\exp\left(\alpha \sup_{T\times T} |X_s - X_t|^\tau\right) < \infty \tag{12}$$

(c) *The non-integrable case.* Let $0<q<p<1$ and let $X:(T,\delta)\to\mathbf{R}$ be a δ-separable random process satisfying one of the two conditions:

$$\sup_{T\times T} \mathbf{E}\left\{\left|\frac{X_s - X_t}{\delta(s,t)}\right|^p\right\} < \infty \tag{13}$$

$$\limsup_{x\to\infty} x^p \sup_{T\times T} P\{|X_s - X_t| > x\delta(s,t)\} < \infty \tag{14}$$

In this case, condition (1) reduces to
there exist $0<q<p$ such that

$$\int_0^{\operatorname{diam}(T,\delta)} N_\delta(T,v)^{q/p} v^{q-1}\,\mathrm{d}v < \infty \tag{15}$$

and we deduce from Theorem 13 that

$$\forall \varepsilon > 0, \qquad \mathbf{E}\,\frac{\sup_{T\times T}|X_s - X_t|^p}{\log \sup_{T\times T}|X_s - X_t|^{1+\varepsilon}} < \infty \tag{16}$$

and

$$\forall \varepsilon > 0 \text{ and for all } x \text{ sufficiently large,}$$

$$P\left\{\sup_{\delta(s,t_0)<\varepsilon} |X_s - X_{t_0}| > \left[\int_0^\varepsilon N(B(t_0,\varepsilon),u)^{q/p}\,u^{q-1}\right]^{1/q} x\,\mathrm{d}u\right\} \leqslant (5q)^{p/q} x^{-p} \tag{17}$$

Proof of Theorem 13 Since ($\mathscr{H}$) is satisfied, (T, δ) is pre-quasi-compact, and therefore separable. Furthermore, the process X is not only δ-separable but also δ-continuous in probability; it is therefore separated by every sequence S dense in (T, δ) and we have

$$\text{almost surely} \quad \sup_{s\in T} X_s - X_{t_0} = \sup_{s\in S} X_s - X_{t_0}$$

We introduce two sequences of positive real numbers:

$$\begin{aligned} &\boldsymbol{\varepsilon} = (\varepsilon_n)_{n\geqslant 1} \text{ decreasing to zero with } \varepsilon_0 = \delta(T) \\ &\mathbf{x} = (x_n)_{n\geqslant 0} \end{aligned} \tag{18}$$

Let S_n denote a minimal sequence of centres of open δ-balls of radius ε_n covering T; we thus have $\#S_n = N_\delta\ (T, \varepsilon_n)$. For each natural number n, we then put $S^n = \bigcup_{j=0}^n S_j, S^\infty = S, S_0 = \{t_0\}$, a point of T, and we use the the notation

$$s \rightarrow \mathbf{s}$$

to denote a mapping of S_n into S_{n-1} defined by the condition $\delta(s, \mathbf{s}) < \varepsilon_{n-1}$.

We shall verify, to begin with, that

$$z_n = \sup_{s\in S^n} |X_s - X_{t_0}| \sum_{k=1}^{n} \varepsilon_{k-1} \sup_{s\in S_k} \left|\frac{X_s - X_{\mathbf{s}}}{\delta(s, \mathbf{s})}\right| \tag{19}$$

Indeed, observe that, for each natural number n,

$$\begin{aligned} 0 \leqslant z_n - z_{n-1} &= \left(\sup_{s\in S^n} |X_s - X_{t_0}| - \sup_{s\in S^{n-1}} |X_s - X_{t_0}|\right) \\ &\leqslant \sup_{s\in S_n} (|X_s - X_{t_0}| - |X_{\mathbf{s}} - X_{t_0}|) \\ &\leqslant \sup_{s\in S_n} |X_s - X_{\mathbf{s}}| \leqslant \varepsilon_{n-1} \sup_{s\in S_n} \left|\frac{X_s - X_{\mathbf{s}}}{\delta(s, \mathbf{s})}\right| \end{aligned}$$

from which, since $z_n = \sum_{p=1}^n z_p - z_{p-1}$, equation (19) follows.

It follows from the fact that $x^+ = \sup(x, 0)$ is increasing and subadditive that

$$\begin{aligned} \left(z_n - \sum_{k=1}^n \varepsilon_{k-1} x_k\right)^+ &\leqslant \left[\sum_{k=1}^n \varepsilon_{k-1} \left(\sup_{s\in S_k} \left|\frac{X_s - X_{\mathbf{s}}}{\delta(s, \mathbf{s})}\right| - x_k\right)\right]^+ \\ &\leqslant \sum_{k=1}^n \varepsilon_{k-1} \left(\sup_{s\in S_k} \left|\frac{X_s - X_{\mathbf{s}}}{\delta(s, \mathbf{s})}\right| - x_k\right)^+ \end{aligned} \tag{20}$$

In these circumstances, we put

$$\forall k \geqslant 0, \qquad \varepsilon_k = 2^{-k} \varepsilon_0$$

$$x_k = \Gamma\left(\frac{1}{N_\delta(T, \varepsilon_k)}\right)$$

and

$$Y_0 = \sum_{k=1}^{\infty} \varepsilon_{k-1}\left(\sup_{S_k}\left|\frac{X_s - X_{\mathbf{s}}}{\delta(s,\mathbf{s})}\right| - x_k\right)^+$$

Now, on the one hand,

$$\begin{aligned}
\mathbf{E}Y_0 &\leqslant \sum_{k=1}^{\infty} \varepsilon_{k-1} N_\delta(T,\varepsilon_k)\int_{x_k}^{\infty} P\left\{\left|\frac{X_s - X_{\mathbf{s}}}{\delta(s,\mathbf{s})}\right| > u\right\} \mathrm{d}u \\
&\leqslant \sum_{k=1}^{\infty} \varepsilon_{k-1} N_\delta(T,\varepsilon_k)\int_{x_k}^{\infty} P\{\Lambda_0 > y\}\,\mathrm{d}u \\
&\leqslant \sum_{k=1}^{\infty} \varepsilon_{k-1} N_\delta(T,\varepsilon_k)\frac{R(N_\delta(T,\varepsilon_k))}{N_\delta(T,\varepsilon_k)} \\
&< \infty, \qquad \text{by hypothesis}
\end{aligned}$$

and on the other hand,

$$\begin{aligned}
\sum_{k=1}^{\infty} \varepsilon_{k-1} x_k &= \sum_{k=1}^{\infty} \varepsilon_{k-1}\Gamma(N_\delta(T,\varepsilon_k)) \\
&= \sum_{k=1}^{\infty} \varepsilon_{k-1} R(N_\delta(T,\varepsilon_{k-1})) \\
&< \infty
\end{aligned}$$

Since, moreover, $(z_n - \sum_{k=1}^{n} \varepsilon_{k-1} x_k)^+$ is almost surely uniformly majorized by $Y_0 \in L^1$, the dominated convergence theorem yields

$$\mathbf{E}\left\{\sup_T(|X(t) - X(t_0)| - \sum_{k=1}^{\infty} \varepsilon_{k-1} x_k)^+\right\} \leqslant \mathbf{E}\{Y_0\} \tag{21}$$

from which the first part of the theorem follows. Putting

$$x_k = \Gamma\left(\frac{a}{N_\delta(T,\varepsilon_k)}\right), \qquad 0 < a < 1,$$

in equation (21), we find that

$$\begin{aligned}
&\mathbf{E}\left\{\left[\sup_T|X_t - X_{t_0}| - \sum_{k=1}^{\infty}\varepsilon_{k-1}\Gamma\left(\frac{a}{N_\delta(T,\varepsilon_k)}\right)\right]^+\right\} \\
&\qquad\leqslant a\sum_{k=1}^{\infty}\varepsilon_{k-1}\left\{R\left(\frac{N_\delta(T,\varepsilon_k)}{a}\right) - \Gamma\left(\frac{a}{N_\delta(T,\varepsilon_k)}\right)\right\}
\end{aligned} \tag{22}$$

However, every r.r.v. U satisfies the following inequality:

$$\forall u \in \mathbf{R},\ \forall \varepsilon > 0, \qquad \mathbf{E}\{(U-u)^+\} \geqslant \varepsilon P\{U > u + \varepsilon\} \tag{23}$$

and putting

$$u = \sum_{k=1}^{\infty} \varepsilon_{k-1} \Gamma\left(\frac{a}{N_\delta(T, \varepsilon_k)}\right)$$

$$\varepsilon = \sum_{k=1}^{\infty} \varepsilon_{k-1} \left[R\left(\frac{N_\delta(T, \varepsilon_k)}{a}\right) - \Gamma\left(\frac{a}{N_\delta(T, \varepsilon_k)}\right) \right]$$

$$U = \sup_{t \in T} |X_t - X_{t_0}|$$

this yields

$$P\left\{\sup_{t \in T} |X_t - X_{t_0}| > \sum_{k=1}^{\infty} \varepsilon_{k-1} R\left(\frac{N_\delta(T, \varepsilon_k)}{a}\right)\right\} \leqslant a \tag{24}$$

from which we get the second part of these theorem.

5.2.2

The preceding results provide conditions that ensure the existence, for the process being studied, of a version with continuous orbits. These express, in terms of entropy, the fact that the metric space (T, δ) must not be too big.

In the same spirit, we here obtain a partial converse. A generalized capacity on (T, δ) is associated with the process under consideration and it is shown that if this capacity is positive, the orbits of X are almost surely unbounded on T. This is a recent (1984) result due to S. M. Berman [1].

Let (T, δ) be a measurable space and consider a measurable random process $X = \{X(\omega, t),\ t \in T,\ \omega \in \Omega\}$ together with a mapping d: $T \times T \to \mathbf{R}^+$ which is measurable, symmetric and vanishes on the diagonal. To start with we assume that (H): the increments $X_s - X_t$ are absolutely continuous; $p(x, s, t)$ denotes their density.

Consider now the following family of hypotheses linking $p(x, s, t)$ with $d(s, t)$.

Hypothesis A Whenever $s \neq t$, $p(x, s, t)$ can be extended analytically into a strip $|\mathrm{Im}\, z| < c, c > 0$, where c is independent of s and t.

Hypothesis B There exists an increasing map h: $\mathbf{R}^+ \to \mathbf{R}^+$ and a number $c > 0$ such that

$$\forall s \neq t, \forall 0 \leqslant x < c, \qquad \int_{-\infty}^{\infty} |p(-\mathrm{i}x + y, s, t)|^2 \,\mathrm{d}y < \frac{1}{d(s,t)} h\left(\frac{x}{d(s,t)}\right)$$

Hypothesis C The densities $x \to p(x, s, t)$ are positive definite; furthermore, there exists a number $c > 0$ such that

$$\forall s \neq t, \forall 0 \leqslant x < c, \qquad |p(-\mathrm{i}x, s, t)| \leqslant \frac{1}{d(s,t)} h\left(\frac{x}{d(s,t)}\right)$$

We then put

$$W_X^c(T) = \inf \int_T \int_T \frac{1}{d(s,t)} h\left(\frac{c}{d(s,t)}\right) \mathrm{d}\mu(s)\,\mathrm{d}\mu(t)$$

where the infimum is taken over all probability measures on $(T, \mathscr{T})$. Under these conditions we have the following theorem:

Theorem 17 (S. M. Berman [1], p. 136) Suppose that hypotheses A and B or A and C are satisfied. For the random process X to be almost surely bounded on T, it is sufficient that:

$$\text{there exists } c > 0 \text{ such that } W_X^c(T) < \infty \tag{25}$$

Remarks In the gaussian case, this condition reduces to:

there exists $c > 0$ and a probability measure μ on $(T, \mathscr{T})$ such that

$$\iint_{T \times T} \exp\left(\frac{c}{\| X_s - X_t \|^2_{L^2(\Omega)}}\right) \mathrm{d}\mu(s)\,\mathrm{d}\mu(t) < \infty \tag{26}$$

This condition is equivalent to the following:

There exists $c > 0$ and a probability measure μ on $(T, \mathscr{T})$ such that

$$\mu \otimes \mu\{(s,t) \in T^2 : \| X_s - X_t \|_{L^2} \leqslant u\} \leqslant \exp(-c/u^2) \tag{27}$$

When X is stationary on $\mathbf{R}$, this condition is satisfied as long as

$$\liminf_{t \to 0} \| X_0 - X_t \|_{L^2} \log \frac{1}{t} > 0$$

Observe that, in this case, condition (25) is a somewhat weak statement.

Now let $X = \{X(\omega, t), 0 \leqslant t \leqslant 1, \omega \in \Omega\}$ be a measurable random process whose increments $X_s - X_t$ are assumed to be absolutely continuous and such that their density $p(x, s, t)$ satisfies

$$\int_{\mathbf{R}} |p(-\mathrm{i}x + y, s, t)|^2 \,\mathrm{d}y < c|\log|s - t\|^{\theta} \exp|bx/\log|s - t|^{\theta}|^{\gamma}$$

where the parameters c, b, γ and $\theta > 0$ are independent of x, s and t.
Theorems 16 and 17 imply

If $\theta\gamma < 1$, then X is almost surely unbounded.
If $\theta\gamma > 1$, then X is almost surely bounded.

The proof of the theorem employs the notion of local time. Let $(T, \mathscr{T}, \mu)$ be a probability space and let $X = \{X(\omega, t),\ t \in T, \omega \in \Omega\}$ be a measurable random process. The stopping distribution of X on T is the image law of the orbit $t \to X(\omega, t)$; to be precise,

$$\forall A \in \mathscr{B}(\mathbf{R}) \text{ and } \forall \omega \in \Omega, \qquad \nu_\omega(A) = \mu\{t \in T : X(\omega, t) \in A\}$$

When this image law is (almost surely) absolutely continuous with respect to the Lebesgue measure λ, we say that the local time of X exists (almost surely). The local time of X will, by definition, be the Radom–Nikodym derivative of ν_ω, i.e.

$$\varphi(\omega,\cdot)=\frac{\mathrm{d}}{\mathrm{d}\lambda}\nu_\omega(\cdot)$$

We notice at once that

$$\hat{\nu}_\omega(\alpha)=\int_{\mathbf{R}}\mathrm{e}^{\mathrm{i}\alpha t}\,\mathrm{d}\nu_\omega(t)=\int_T \mathrm{e}^{\mathrm{i}\alpha X(\omega,t)}\,\mathrm{d}\mu(t)$$

This relation enables us to check easily when $\hat{\nu}_\omega$ is almost surely square integrable, in which case φ exists and is square integrable. In the gaussian case, one thus obtains

$$\left\{\begin{array}{l}\text{The local time of}\\ X \text{ exists a.s. and}\\ \text{belongs to } L^2(\mathbf{R})\end{array}\right\}\Leftrightarrow\left\{\int_T\int_T\frac{\mathrm{d}\mu(s)\,\mathrm{d}\mu(t)}{\|X_s-X_t\|_{L^2}}<\infty\right\}\tag{28}$$

When, further, the measure ν_ω is equivalent to λ, the orbits of X spend a non-zero amount of time in every measurable set A of non-zero μ-measure; they are therefore unbounded. This is the situation when the local time of X can be extended analytically into a strip $|\mathrm{Im}\, z|<c>0$. This is an idea which is to be found in the proof of the theorem of S. M. Berman, which we give in outline here.

Proof of Theorem 17 By hypothesis

$$\int_{T\times T}\int_{\mathbf{R}}|p(-\mathrm{i}x+y,s,t)|^2\,\mathrm{d}y\,\mathrm{d}\mu(s)\,\mathrm{d}\mu(t)<\infty$$

Moreover, $p(x,s,t)$ can be extended analytically into the strip $|\mathrm{Im}\, z|<c$ and we have

$$\forall\varepsilon>0,\qquad \sup_{0\leqslant x\leqslant c-\varepsilon}\int|p(-\mathrm{i}x+y,s,t)|^2\,\mathrm{d}y\leqslant\frac{1}{d(s,t)}\,h\left(\frac{c-\varepsilon}{d(s,t)}\right)$$

By a theorem due to R.E.A.C. Paley and N. Wiener ([35], p. 7), there exists a measurable function f such that

$$f\in L^2(\mathbf{R})\qquad\text{and}\qquad\int_{\mathbf{R}}|f(x)|^2\,\mathrm{e}^{2(c-\varepsilon)x}\,\mathrm{d}x<\infty$$

and, further,

$$\forall\, 0\leqslant x\leqslant c-\varepsilon,$$

$$\lim_{A\to\infty}\int_{-\infty}^{\infty}|p(-\mathrm{i}x+y,s,t)\int_{-A}^{A}f(u)\,\mathrm{e}^{-u(-x+\mathrm{i}y)}\,\mathrm{d}u|^2\,\mathrm{d}y=0$$

Hence,

$$p(-\mathrm{i}x+y,s,t)=\int_{-\infty}^{+\infty} f(u)\,\mathrm{e}^{-ux+\mathrm{i}yu}\,\mathrm{d}u, \qquad y \text{ almost everywhere,}$$

so by Plancherel's theorem

$$f(u)=\hat{p}(u,s,t), \qquad u \text{ almost everywhere}$$

and the Parseval relation implies

$$\int_{-\infty}^{\infty}|p(-\mathrm{i}x+y,s,t)|^2\,\mathrm{d}y=\int_{-\infty}^{\infty}|\hat{p}(u,s,t)|^2\,\mathrm{e}^{2ux}\,\mathrm{d}x$$

We have therefore shown that

$$\int_{T\times T}\int_{\mathbf{R}}|\hat{p}(u,s,t)|^2\,\mathrm{e}^{2ux}\,\mathrm{d}u\,\mathrm{d}\mu(s)\,\mathrm{d}\mu(t)<\infty$$

Now $\hat{p}(u,s,t)=\mathbf{E}\{\exp[\mathrm{i}u\,X(t)-X(s)]\}$. Choose $\varepsilon=c/2$ and then x' with $x<x'<c/2$. By the Cauchy–Schwarz inequality

$$\left[\int_{T\times T}\int_0^{\infty}|\hat{p}(u,s,t)|\,\mathrm{e}^{2ux}\,\mathrm{d}u\,\mathrm{d}\mu(s)\,\mathrm{d}\mu(t)\right]^2$$

$$\leqslant\int_0^{\infty}\left[\iint_{T\times T}|\hat{p}(u,s,t)|^2\,\mathrm{d}\mu(s)\,\mathrm{d}\mu(t)\right]^2\mathrm{e}^{4x'u}\,\mathrm{d}u\int_0^{\infty}\mathrm{e}^{-4(x'-x)u}\,\mathrm{d}u$$

$$\leqslant\frac{1}{4(x'-x)}\int_0^{\infty}\int_{T\times T}|p(u,s,t)|^2\,\mathrm{d}\mu(s)\,\mathrm{d}\mu(t)\,\mathrm{e}^{4x'u}\,\mathrm{d}u$$

$$<\infty$$

Since, moreover, the function $|p(u,s,t)|$ is even, being the modulus of a characteristic function, we have shown that, if $0<x<c/2$,

$$\int_{\mathbf{R}}\int_{T\times T}|\hat{p}(u,s,t)|\,\mathrm{d}\mu(s)\,\mathrm{d}\mu(t)\,\mathrm{e}^{2xu}\,\mathrm{d}u=\mathbf{E}\left\{\int_{\mathbf{R}}\left|\int_T \mathrm{e}^{\mathrm{i}uX(s)}\,\mathrm{d}\mu(s)\right|^2\mathrm{e}^{2xu}\,\mathrm{d}u<\infty\right.$$

From this we deduce, by a well-known theorem (see, for example, [24], p. 439, theorem II.7.1), that the local time of X is analytic in the strip $|\mathrm{IM}\,z|<c/2$, whence the conclusion of Theorem 17.

5.2.3 Ultrametric structures

Let (T,δ) be a metric space. When δ satisfies the 'strong' triangle inequality

$$\forall\, s,t,u\in T, \qquad \delta(s,t)\leqslant\max(\delta(s,u),\delta(u,t)) \tag{29}$$

the metric is said to be non-archimedean or ultrametric, and (T,δ) is then called an ultrametric space.

The main feature o these spaces lies in the fact that two balls of the same radius are either disjoint or the same, so that

$$\forall \varepsilon > 0, \qquad M_\delta(T, \varepsilon) = N_\delta(T, \varepsilon)$$

Now suppose that (T, δ) is a separable ultrametric space. It is easy to show that (T, δ) can be embedded continuously in a projective limit of sets with compact ultrametric structure. One proceeds in the following way. Consider, for each natural number n, the following mappings:

(a) θ_n: $T \to S_n$ defined by $\delta(s, \theta_n(s)) \leqslant \varepsilon_n$

   where $S_n \subset T$ is a minimal sequence of centres of closed balls of radius $\varepsilon_n = 2^{-n}\,\mathrm{diam}(T, \delta)$.

(b) $\Pi_{n,n-1}$: $S_n \to S_{n-1}$ defined by $\delta(t, \Pi_{n,n-1}(t)) \leqslant \varepsilon_{n-1}$.

(c) For each natural number k with $k \leqslant n$,

$$\Pi_{n,k} = \bigotimes_{l=k=1}^{n} \Pi_{l,l+1} : S_n \to S_k$$

   with the convention $\Pi_{n,n} = \mathrm{Id}(S_n)$.

The following lemma is almost immediate.

Lemma The pair $((S_n), \Pi_{n,k}))$ forms a projective system of sets and we have

$$\theta_k = \Pi_{n,k} \circ \theta_n, \; n \geqslant k \geqslant 0$$

We write $L = \varprojlim\, (S_n, \Pi_{n,k})$, its projective limit, $G = \prod_1^\infty S_n$, and, for each integer $k \geqslant 0$, let Π_k be the restriction to L of the projection of G onto S_k. We also put, for any elements s, t of L,

$$\Delta(s, t) = \varepsilon_{n(s,t)}$$

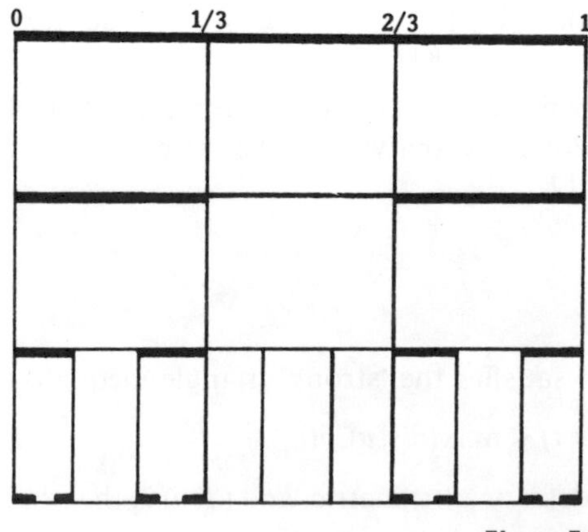

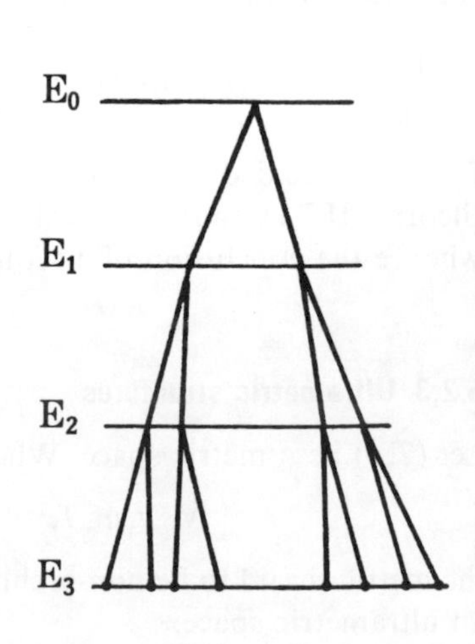

Figure 5.4

where $n(s,t)=\sup\{n\geqslant 0\colon \Pi_n(s)=\Pi_n(t)\}$. Thus, (L,Δ) is a compact ultrametric space. Furthermore, the mapping $l\colon (T,\delta)\to(L,\Delta)$ defined by $l(s)=\{\theta_n(s), n\geqslant 0\}$ determines a continuous embedding of (T,δ) into (L,Δ) and we have the relations

$$\forall s,t\in T, \qquad \tfrac{1}{2}\Delta(l(s),l(t))\leqslant\delta(s,t)\leqslant\Delta(l(s),l(t))$$

Finally, when (T,δ) is itself compact, l is an isomorphism with respect to uniform structures.

Figure 5.4 provides a good illustration of the situation, in which the construction of the Cantor set is the chosen example.

$$E=(x_n)\in 0,1,2; \qquad x_n=0 \text{ or } 2\ \forall n$$

The interest of these spaces lies in the fact that the search for necessary and sufficient conditions for the continuity of a gaussian process indexed on a space of this type (with $\delta(s,t)=\|X_t-X_s\|_{L^1}$) has turned out to be relatively straightforward, with the follwoing characterization due to X Fernique.

Theorem 18 ([10]) Let $X=\{X(\omega,t), t\in T, \omega\in\Omega\}$ be a centred gaussian process indexed by T; put

$$\delta(s,t)=\|X_s-X_t\|_{L^1(\Omega)}$$

Suppose that (T,δ) is a compact ultrametric space. For the orbits of X to be almost surely majorized, it is necessary and sufficient that the following condition be satisfied:

$$\sup_\mu \int_T d\mu(t)\int_0^{\operatorname{diam}(T,\delta)}\sqrt{\log\frac{1}{\mu(s\in T\colon \delta(s,t)<u)}}\,du<\infty$$

where the maximum is taken over the set of all Borel probability measures on T.

This theorem is proved by considering an auxiliary gaussian process indexed by (T,δ), defined, starting with a normal gaussian sequence $(\Lambda_n, n\in\sum S_n)$, as follows:

$$\forall t\in T, \qquad Z(t)=\sum_{n=0}^{\infty}\Lambda_{\pi n(t)}\operatorname{diam}(T,\delta)\times 2^{-n}$$

It is easily checked that

$$d_Z(s,t)=\|Z_s-Z_t\|_{L^2}=2^{-n(s,t)}\operatorname{diam}(T,\delta)\sqrt{\tfrac{3}{2}}$$

whence

$$\sqrt{\tfrac{3}{2}}\,\delta(s,t)\leqslant d_Z(s,t)\leqslant\sqrt{6}\,\delta(s,t)$$

The result follows from this, onece one has characterized the almost sure majorization of the process Z, using the comparison lemma of Slepian–Fernique ([91], theorem II.1.2).

References and bibliography

1. Berman S. M. (1984) Unboundedness of sample functions of stochastic processes with arbitrary parameter sets, with applications to linear and L_p-valued parameters, *Osaka J. Math.*, **21**, 133–147.
2. Bayer W. A. (1962) Hausdorff dimensions of level sets of some Rademacher series, *Pacif. J. Math.*, **12**, 35–46.
3. Bouligand G. (1928) Ensembles impropres et ordre dimensionnel, *Bull. Sci. Math.*, **52**, 320–344 and 361–376.
4. Ciesielski Z. and Taylor S. J. (1962) First passage times and sojourn times for brownian motion in space and the exact Hausdorff measure of the sample path, *Trans. Am. Math. Soc.*, **1**, 434–450.
5. Dvoretsky A. (1948) A note on Hausdorff dimension functions, *Proc. Camb. Phil. Soc.*, **44**, 13–16.
6. Dvoretsky A., Erdös P. and Kakutani S. (1954) Double points of paths of brownian motion in *n*-spaces, *Acta Sci. Math. Szeged*, **12**, 75–81.
7. Dvoretsky A., Erdös P. and Kakutani S. Multiple points of paths of brownian motion. *Bull. Res. Counc., Israel*, **3**, 364–372.
8. Dvoretsky A., Erdös P., Kakutani S. and Taylor S. J. Triple points of brownian motion in 3-space, *Proc. Camb. Phil. Soc.*, **53**, 856–862.
9. Fernique X. (1975) Régularité des trajectoires de fonctions aléatoires gaussiennes, Ecole d'Eté de St Flour, Lecture Notes 480.
10. Fernique X. (1977) Caractérisation de processus a trajectoires presque sûrement majorées ou continues, Séminaire de Prob. XII, Univ. Strasbourg, Lecture Notes 649.
11. Fernique X. (1981) Régularité de fonctions aléatoires non gaussiennes, Ecole d'Eté, St Flour, Lecture Notes 976.
12. Frostman O. (1935) Thesis, Meddelanden Lunds Univ., Math. Sem. 3.
13. Goldman A. (1981) Points multiples de processus gaussiens. *Z. Wahrscheinlichtkeitsth.*, **34**, 481–494.
14. Hahn M. G. (1975) Central limit theorem for $D[0,1]$ valued random variables, Ph.D. Thesis, MIT, Cambridge, Mass.
15. Hausdorff F. (1919) Dimension und äusseres Mass, *Math. Ann.*, **79**, 159–179.
16. Hurewicz W. and Wallman H. (1941) *Dimension theory*, Princeton Univ. Press, Princeton.
17. Ibragimov I. (1979) Sur la régularité des trajectoires des fonctions aléatoires, *C.R. Acad. Sci., Paris*, **A289**, 545–547.
18. Jarnik V. Diaphantisches Approximationen und Hausdorffsches Mass, *Recueil Math. de la Société de Moscou*, **36**, 371–382.
19. Jernik V. (1931) Uber die simultanen diaphantischen Approximationen, *Math. Zeitschrift*, **33**, 503–543.
20. Kaczmarz S. and Steinhaus H. (1930) Le Système orthogonal de M. Rademacher, *Studia Math.*, **2**, 231–247.
21. Kahane J.-P. (1983) Ensembles aléatoires et dimensions, Cours Prépubl., Univ. Orsay.
22. Kahane J.-P. and Salem R. (1963) *Ensembles Parfaits et Séries Trigonométriques*, Hermann, Paris.

23. Kametani S. (1944) On Hausdorff measures and generalised capacities with some applications to the theory of functions, *Jap. T. Math.*, **19**, 217–257.
24. Kaufman R. (1976) Covering properties of random series, *Duke Math. J.*, **43**(2) 295–299.
25. Kawada (1972) *Fourier Analysis in Probability Theory*, Academic Press.
26. Kolmogorov A. N. and Tihomirov V. M. (1959–61) Epsilon entropy and epsilon capacity in functional spaces, *Uspehi Mat. Nauk.*, **14**(2), 3–86. Translation in *Am. Math. Soc. Trans.*, Series 2, **17**, 277–364.
27. Kôno N. (1978) Best possibility of an integral test for sample continuity of L^p-process ($p > 2$), *Proc. Jap. Acad.*, **54**, Series 1, 197–201.
28. Loeve M. (1963) *Probability Theory*, 3rd ed., Van Nostrand.
29. Mandelbrot B. and Van Ness J. W. (1966) Fractional noises and applications, *Siam. Rev.*, **10**, 422–437.
30. Marstrand J. M. (1953) Some fundamental geometrical properties of plane sets of fractional dimensions, *Proc. Lond. Math. Soc.*, **4**, 257–302.
31. Marstrand J. M. (1954) The dimension of the cartesian product sets, *Proc. Camb. Phil. Soc.*, **50**, 198–202.
32. Munroe M. E. *Measure and Integration*, 2nd ed., Addision-Wesley.
33. Nagata J. (1983) *Modern Dimension Theory*, Revised ed., Helderman Verlag, Berlin.
34. Nanopoulos C. and Nobelis P. (1977) Etude de la régularité des fonctions alćatoires et leurs propriétés limites, Thesis, Publ., Strasbourg.
35. Paley R.E . A. C. and Wiener N. (1934) Fourier transform in the complex domain, Am. Math. Soc. Colloq. Publ. 19, Am. Math. Soc., New York.
36. Pisier G. Conditions d'entropie assurant la continuité de certains processus..., Sem. Ann. Fonct. Exp. 13–14, Ec. Polytech., Palaiseau.
37. Pontjragin L. and Schnirelman L. (1983) Sur une propriété métrique de la dimension, *Ann. Math.*, **33**, 156–162.
38. Rogers C. A. (1970) *Hausdorff Measures*, Cambridge Univ. Press.
39. Rogers C. A. and Taylor S. J. (1961) functions continuous and singular with respect to a Hausdorff measure, *Mathematika*, **8**(15), 1–32.
40. Taylor S. J. (1953) On cartesian product sets, *J. Lond. Math. Soc.*, **27**, 295–304.
41. Taylor S. J. (1964) On connexion between Hausdorff measure and generalized capacity. Proc. Camb. Phil. Soc., **57**, 524–531.
42. Weber M. (1980) Analyse asymptotique de processus gaussiens stationnaires, *Ann. Inst. H. poincaré*, **XVI**(2), 117–176.
43. Weber M. (1981) Une méthode élémentaire pour l'étude de la régularité d'une large classe de fonctions aléatoires, *C.R. Acad. Sci., Paris*, **292**, 599–602.
44. Weber M. (1981) Analyse infinitésimale de fonctions aléatoires, Ecole d'Eté Prob., St Flour, Lecture Notes in Math. 976, Springer.
45. Weber M. (1983) Dimension de Hausdorff et points multiples du mouvement brownien fractionnaire dans $\mathbf{R}^n$, *C.R. Acad. Sci., Paris*, **297**, Series I, 357–360.

Chapter 6 Attractors and dimensions

Paul Girault
Department of Mathematics, University of Strasbourg

6.1 Introduction

The last twenty or so years have seen a quite significant increase in the amount of work devoted to the study of non-linear or turbulent phenomena whose common feature is seemingly chaotic time evolution: hydrodynamics, population studies, electric circuits and chemical kinetics all come to mind. One can identify various reasons for this, among others, the fact that certain experimental data have only recently become available and, above all, the growing importance of numerical modelling thanks to the ever more widespread use of electronic computers. It will be noticed, however, that the capacity of such machines plays but a modest role in this—just remember Feigenbaum's discovery. Finally, this improved understanding has only been possible thanks to an effective and vigorously developing theory, namely the qualitative theory of differentiable dynamical systems. This theory has experienced several breakthroughs, some very recent: one thinks of bifurcation theory, ergodic theory, the introduction of the concept of strange attractors, as well as everything to do with notions of dimension, Lyapounov exponents, etc. This is a suitable time once more to recollect that the concept of strange attractor would have been inconceivable in the precomputer age: see the works of, on the one hand, E. N. Lorenz [21] and M. Hénon [14] on the other. Beyond the fact that numerical modelling has revived interest in certain questions and created new problems, it is nevertheless undeniable that the strange and fascinating shapes and patterns associated with this work have stimulated researchers—not only smokers dream before the plumes of Von Karman! The numerous illustrations in the work of I. Gumowski and C. Mira [13], entitled *Chaotic dynamics*, are a particularly telling sign of a new state of mind. This contrasts with the pious and respectful oblivion into which the nonetheless remarkable works of Fatou and Julia have fallen. For an elementary survey of the subject of chaotic dynamics, see the readable and well-illustrated book of P. Berge *et al.* [4].

In the multidisciplinary setting of this seminar, the reader is invited to take a stroll among some representative work on the notions of attractors—of their dimensions as well as their physical interpretation when this is possible.

6.2 Some definitions

6.2.1 Hausdorff measure and dimension

Let X be a metric space (usually a Hilbert space in what follows) and let $Y \subset X$; for $d \varepsilon \mathbf{R}$ and $\varepsilon > 0$, put

$$\mu(Y, d, \varepsilon) = \inf \sum_i r_i^d$$

where the infimum is taken over all coverings of Y by a family of balls $(B_i)_{i \varepsilon I}$ of X, of radii $r_i \leqslant \varepsilon$.

The function $\varepsilon \to \mu(Y, d, \varepsilon)$ is decreasing and therefore has a limit $\mu(Y, d) \varepsilon \bar{\mathbf{R}}_+$ as $\varepsilon \to 0$, so that

$$\mu(Y, d) = \lim_{\varepsilon \to 0} \mu(Y, d, \varepsilon) = \sup_{\varepsilon > 0} \mu(Y, d, \varepsilon)$$

This number is the d-dimensional *Hausdorff measure* of Y. Starting with this, one can show that there exists a real number $d_0 \varepsilon \bar{\mathbf{R}}_+$ such that $\mu(Y, d) = \infty$ for $d < d_0$ and $\mu(Y, d) = 0$ for $d > d_0$. This real number $d_0 = d_H(Y)$ is called the *Hausdorff dimension* of Y.

6.2.2 Fractal dimension

Let X be a metric space (generally a Hilbert space here) and let $Y \subset X$.

If $N_Y(\varepsilon)$ is the least number of balls of radius less than ε which are needed to cover Y, then the number $d_F(Y)$ defined by

$$d_F(Y) = \limsup_{\varepsilon \to 0} \frac{\ln N_Y(\varepsilon)}{\ln 1/\varepsilon}$$

is the fractal dimension of Y (sometimes called the capacity of Y [23, 32], the Mandelbrot dimension or the Schnirelman–Kolmogorov dimension of Y). It is well known that

$$d_F(Y) = \inf \{ d > 0, \limsup_{\varepsilon \to 0} \varepsilon^d N_Y(\varepsilon) = 0 \}$$

(see [22]). We always have $d_H(X) \leqslant d_F(X)$.

For all these matters, see B. Mandelbrot [22] or the article of M. Weber [30] in this seminar.

Remark In section 6.3, we will introduce a new dimension, the Lyapounov dimension (see also [11]).

6.2.3 The notion of attractors

In the qualitative study of differentiable dynamical systems, the study of attractors has particularly attracted the attention of theorists because of their complicated geometrical structure. Unfortunately, there is a lack of agreement on how they should normally be defined, although we have to thank J. Milnor for bringing some order to the matter in a recent article [25].

We begin by recalling one of the most generally used definitions in the qualitative theory of differential equations. Let X be a vector field, $\{\varphi_t\}_{t\in\mathbf{R}}$, the flow associated with X. A compact set A invariant under the action of the flow (i.e. $\varphi_t(A) = A, \forall t$) is an *attractor* (or asymptotically stable) if there exists a neighbourhood V of A such that

$$\bigcap_{t\geqslant 0} \varphi_t(V) = A$$

This definition is taken from [24]. Some authors add further conditions, according to taste. More suggestively, we shall say, following C. Sparrow [29], that A is an attractor if A is invariant under the action of the flow and if, for each x sufficiently close to A, the orbit passing through x tends to A. Thus, an attractor A is the asymptotic limit of the solutions setting out from those initial conditions lying in a set B (clearly $A \subset B$), called the *basin of attraction* of A (see also [4]).

In the theory of point transformations, there is an analogous definition in [13], shown in Figure 6.1. One will observe that the geometrical structure of A and B is very complicated in general; this is also true of that of the frontier of B which, in the case of plane point-transformations, the study of which is more developed, sometimes behaves like a separator.

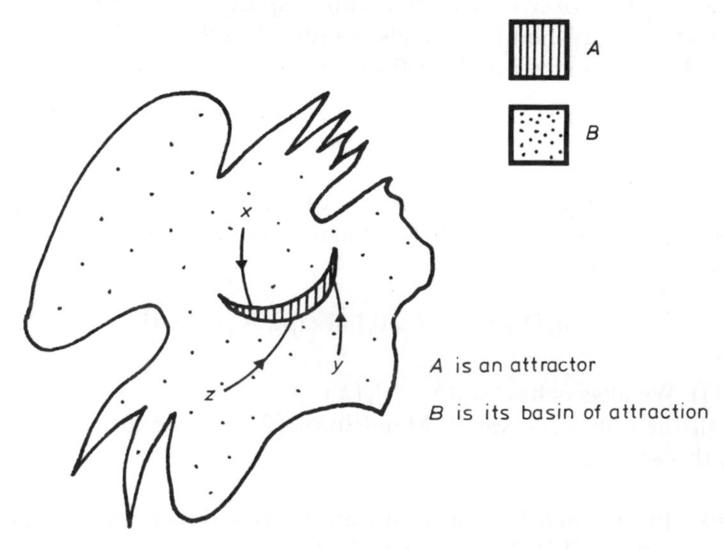

Figure 6.1

Here is an example with a simple frontier; consider the following recurrence formulae:

$$x_{n+1} = y_n$$
$$y_{n+1} = 0.5x_n + 0.1x_n^2$$

The origin O of the (x, y) plane is a fixed attractive point. The frontier F of the basin of attraction of this fixed point is the square ABCD whose corners are A(5, 5), B(−10, 5), C(−10, −10) and D(5, −10). (For further details, see [13].) It will be seen that although the frontier may be simple, it is not differentiable at all points, even though the recurrence has analytic second terms. For more on these problems, see [13], which, further has a very full bibliography as well as excellent diagrams.

The reader who wants to know about other definitions of attractors is referred to the aforementioned article of Milnor [25].

Milnor's main criticism is that most of the definitions are too restrictive and rule out many interesting examples; furthermore, they sometimes lead to an unsatisfactory state of affairs in which nothing has been gained as far as describing either the destination of the majority of points under the action of a flow or the iterates of a mapping is concerned. In the article [25], Milnor has given a definition that is wide enough for a smooth compact dynamical system to admit at least one attractor. Here is his definition.

Let M be a smooth, compact variety, possibly with boundary, and f a continuous mapping of M into itself. The *ω-limit* set $\omega(x)$ of the point $x \varepsilon M$ is, by definition, the set of accumulation points of the sequence $x, f(x), f^2(x), \ldots$. A measure on M is chosen whose restriction to each coordinate neighbourhood is equivalent to Lebesgue measure.

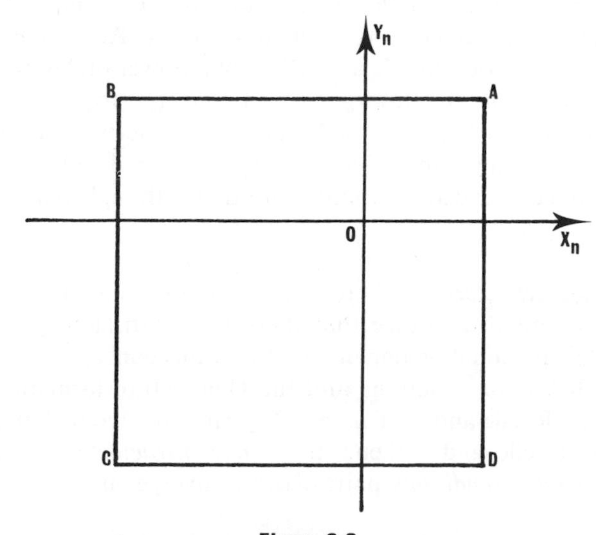

Figure 6.2

Definition A closed subset $A \subset M$ is an *attractor* if it satisfies the following two conditions:

(a) The *domain of attraction* $\rho(A) = \{x \varepsilon M : \omega(x) \subset A\}$ has strictly positive measure.
(b) There is no closed subset A' strictly contained in A such that $\rho(A') = \rho(A)$ modulo a set of measure zero.

Lemma If A is a compact subset of positive measure such that $f(A) \subset A$, then A contains at least one attractor.

Since the proof is simple, we sketch it! Let A_1 be the smallest closed subset of M which contains $\omega(x)$ for almost every x of A. It is easy to see that A_1 is an attractor, in Milnor's sense, of course. Note that its domain necessarily contains A modulo a set of measure zero. It is clear that $A_1 \subset A$.

Example This is the time to give the example of an attractor in Milnor's sense which is not, however, an attractor under most of the usual definitions.

Let $c = 1.401155189\ldots$, the smallest real number for which the mapping $x \rightarrow x^2 - c$ has infinitely many distinct orbits. This mapping has been studied by Feigenbaum. Put $f(x) = x^2 - c$ and let I be an interval containing the origin and such that $f(I) \subset I$, so that f can be regarded as a mapping of I into itself. The orbit of the origin under f is an almost periodic sequence which can be described in the following way: the numbers $0 = a_0 \rightarrow a_1 \rightarrow a_2 \cdots$ are all distinct; furthermore, the difference $a_m - a_n$ is very small when $m - n$ is divisible by a high power of 2. The closure A of this orbit is thus a Cantor set homeomorphic to the ring $\underleftarrow{\lim}(\mathscr{Z}/2^k\mathscr{Z})$ of dyadic integers in such a way that to each $a_n = f^n(0)$ corresponds the dyadic integer n. Observe that f restricted to A corresponds to the homeomorphism which adds 1 to each dyadic integer. It follows that there is no periodic point in A; on the other hand, one can show that there are periodic points arbitrarily close to each point of A. All of these points are unstable and have a period that is a multiple of a power of 2. For almost every initial point x_0 of I, the iterates $x_n = f^n(x_0)$ converge to A, so that A is the unique attractor of f, in the sense of Milnor. Nevertheless, one can show that A is not asymptotically stable! See [25] for further details. Thus, Milnor's definition does not demand any stability condition, although many attractors are asymptotically stable.

Appendix: strange attractors If there is no consensus on the definition of an attractor, it is essential to realize that there is no satisfactory definition of a strange attractor. It should be remembered that this concept was introduced in connection with Lorenz equations and the Henon transformation. In a now-famous work, D. Ruelle and F. Takens [28] carefully avoided giving a precise definition. For Ruelle and Takens, a *strange attractor* is characterized by *sensitivity to the initial conditions*: pairs of orbits diverge on the attractor (see also [26]).

Some authors, such as, for example, Berge *et al.* in [4], impose an additional condition of a dimensional kind: specifically, an attractor is strange if it is sensitive to the initial conditions and if, further, it has non-integral dimension. Of course, not all attractors are strange; consider attractive fixed points or limit cycles. In view of the classical examples (see later), it is natural to ask if there is any link between geometric irregularity and strange dynamical behaviour; in fact, there is no such link as the foregoing example shows. (For other examples, see Milnor's article [25], where this matter is raised and a start is made in the discussion of it.)

6.3 Some theorems on the dimension of attractors

In this chapter, we give several theorems concerning the Hausdorff dimension and the fractal dimension of an attractor, in the general sense of the term. Of course, the calculation of a dimension is often difficult, and one can really only hope to find bounds, which, however, may well suggest things. Results along these lines have been obtained by A. Douady and J. Oesterle [8], P. Constantin and C. Foias [5] and P. Constantin *et al.* [6]. Here, we shall mainly follow [6], because these results are the most interesting from the point of view of applications.

Let H be a Hilbert space, $X \subset H$ a compact set and S a continuous mapping of X into H such that

$$SX = X \tag{1}$$

Suppose further that, $\forall u \in X$, there exists a compact linear operator on H, $L(u) \in \mathscr{L}(H, H)$, such that

$$\sup_{u,v} \frac{\| Sv - Su - L(u)(v-u) \|_H}{\| v - u \|_H} \longrightarrow 0, \qquad 0 < \| u - v \|_H \leqslant \varepsilon, \text{ as } \varepsilon \to 0 \tag{2}$$

Thus S satisfies a uniform differentiability condition. If L is a compact linear operator on H, $B(0, 1)$ denotes the ellipsoid in H with semi-axes $\alpha_j(L)$, $j \geqslant 1$, the eigenvalues of $(L^* L)^{1/2}$, so that $\alpha_1(L) \geqslant \alpha_2(L) \geqslant \cdots$.

For d a natural number, we put

$$\omega_d(L) = \alpha_1(L) \cdots \alpha_d(L)$$

so that $\omega_d(L) = \| \Lambda^d L \|$, where Λ^d is the dth exterior power of the operator L.

If $d \in R_+$, we can write $d = n + s$, where $0 < s < 1$ and n is a natural number, and we put

$$\omega_d(L) = \omega_n(L)^{1-s} \omega_{n+1}^s(L) = \alpha_1(L) \cdots \alpha_n(L) [\omega_{n+1}(L)]^s$$

Since the mapping $d \to [\omega_d(L)]^{1/d}$ is decreasing, $[\omega_d(L)]^{1/d}$ tends to a limit in the interval $[0, \alpha_1(L)]$ as $d \to +\infty$.

We make the following additional hypotheses regarding $L(u)$:

$$\sup_{u \in X} \| L(u) \|_{\mathscr{L}(H,H)} < +\infty \tag{3}$$

and there exists $d > 0$ such that $\sup_{u \in X} \omega_D(L(u)) < 1$ (4)

Theorem 1 Under hypotheses (1), (2), (3) and (4), the Hausdorff dimension of X is finite and, furthermore,

$$d_H(X) \leqslant d$$

For results of this kind, see [8], and, more generally, [23] and [17]. As far as the proof is concerned, see [6].

For the fractal dimension, we have a somewhat less precise result. Let

$$\bar{\omega}_1 = \sup_{u \in X} \omega_1(L(u))$$

and introduce the following hypothesis:

There exists $d = n + s$, with n a natural number and $0 < s < 1$ such that

$$\bar{\omega}_{n+1}^{(d-1)/(n+1)} \bar{\omega}_l < 1 \qquad \text{for } l = 1, 2, \ldots, n \tag{4a}$$

Theorem 2 Under hypotheses (1), (2), (3) and (4a), the fractal dimension of X is finite and, furthermore,

$$d_F(X) \leqslant d$$

It should be noted that there are only a few results on the fractal dimension. (Nevertheless, see [23]. For the proof, see [6].)

Bringing in the Lyapounov exponents and numbers, Constantin *et al.* [6] have obtained a different and very interesting version of Theorems 1 and 2. To begin with, let us look at some notation. $L_p(u)$ will stand for the expression

$$L_p(u) = L(S^{p-1}(u)) \circ \cdots \circ L(S(u)) \circ L(u)$$

Let $\bar{\omega}_j(p) = \sup_{u \in X} \omega_j(L_p(u))$ $(\leqslant \bar{\omega}_j(p))$. It is then easy to show that $\bar{\omega}_j(p+q) \leqslant \bar{\omega}_j(p)\ \omega_j(q)$, $\forall (p, q, j) \in \mathbf{N}^3$. The mapping $p \rightarrow [\bar{\omega}_j(p)]^{1/p}$ thus has a limit as $p \rightarrow +\infty$, denoted by Π_j, $\forall j \geqslant 1$, so that

$$\Pi_j = \lim_{p \rightarrow +\infty} [\omega_j(p)]^{1/p} = \inf_{p > 0} [\omega_j(p)]^{1/p}$$

Letting $\Lambda_1, \ldots, \Lambda_m, \ldots$ be the sequence defined by

$$\Lambda_1 = \Pi_1$$
$$\Lambda_1 \Lambda_2 = \Pi_2$$
$$\cdots\cdots\cdots\cdots\cdots\cdots$$
$$\Lambda_1 \Lambda_2 \cdots \Lambda_j = \Pi_j$$

we then have that, for $j \geqslant 2$,

$$\Lambda_j = \lim_{p \rightarrow +\infty} \left[\frac{\bar{\omega}_j(p)}{\bar{\omega}_{j-1}(p)} \right]^{1/p}$$

The real numbers Λ_j are the *uniform Lyapounov numbers*. The *uniform Lyapounov exponents* are thus defined by

$$\mu_j = \ln \Lambda_j, \qquad j \geqslant 1$$

One also defines the local numbers

$$\Lambda_j(u) = \lim \sup \{\alpha_j(L_p(u))\}^{1/p}$$

$$\mu_j(u) = \ln \Lambda_j(u), \qquad \forall u \in X, \forall j \geqslant 1$$

Finally, put

$$\bar{\Lambda}_j = \limsup_{p \to \infty} \left\{ \sup_{u \in X} \alpha_j(L_p(u)) \right\}^{1/p}$$

$$\bar{\mu}_j = \ln \bar{\Lambda}_j, \qquad \forall j \geqslant 1$$

It is easy to deduce from the above that

$$\Lambda_j \leqslant \bar{\Lambda}_j \leqslant (\Lambda_1 \cdots \Lambda_j)^{1/j}$$

$$\mu_j \leqslant \bar{\mu}_j \leqslant \frac{\mu_1 + \cdots + \mu_j}{j}, \qquad \forall j \geqslant 1$$

Similarly, if $d = n + s$, where $n \in \mathbf{N}$, $0 < s < 1$, is not a whole number, we put

$$\omega_d(p) = \sup_{u \in X} \omega_d(L_p(u))$$

One can show that $\omega_d(p)^{1/p}$ has a limit as $p \to \infty$, so that Π_d may be defined by

$$\Pi_d = \lim_{p \to \infty} (\bar{\omega}_d(p))^{1/p} = \inf_{p > 0} [\bar{\omega}_d(p)]^{1/p}$$

In [6], Constantin *et al.* prove the following theorem:

Fundamental theorem Under hypotheses (1), (2) and (3), if there exists a natural number n such that

$$\mu_1 + \cdots + \mu_{n+1} < 0$$

then

$$\mu_{n+1} < 0, \qquad \frac{\mu_1 + \cdots + \mu_n}{|\mu_{n+1}|} < 1$$

and, further,

(a) We have $d_H(X) \leqslant n + \left| \dfrac{\mu_1 + \cdots + \mu_n}{|\mu_{n+1}|} \right|$

(b) We have $d_F(X) \leqslant \max\limits_{1 \leqslant l \leqslant n} \left(n + \dfrac{\mu_1 + \cdots + \mu_l}{|\bar{\mu}_{n+1}|} \right)$

Two important remarks are in order:

Remark 1 The best value of n in (a) is obained when n is the smallest integer such that $\mu_1 + \cdots + \mu_n \geqslant 0$ and $\mu_1 + \cdots + \mu_{n+1} < 0$, in which case,

$$\frac{\mu_1 + \cdots + \mu_n}{|\mu_{n+1}|} \in (0, 1)$$

For this value of n, the number $n + (\mu_1 + \cdots + \mu_{n+1})/|\mu_{n+1}|$ is the *Liapounov dimension* of X and will be denoted by $d_L(X)$. Thus, $d_H(X) \leqslant d_L(X)$.

In fact, it was conjectured by Frederickson *et al.* [11] that $d_F(X) = d_L(X)$, but following discussions with L. S. Young, the authors are convinced that the strict inequality $d_L(X) < d_F(X)$ can hold (see also [31] and [32]).

Remark 2 In fact, it is easy to see that

$$d_F(X) \leqslant \max_{1 \leqslant l \leqslant n} \left(n + \frac{\mu_1 + \cdots + \mu_l}{|\bar{\mu}_{n+1}|} \right)$$
$$\leqslant (n+1) \max_{1 \leqslant l \leqslant n} \left(1 + \frac{|\mu_1 + \cdots + \mu_l|}{|\mu_1 + \cdots + \mu_{n+1}|} \right)$$

For the proof, we refer the reader to [6].

The fundamental theorem will be our principal tool in what follows.

6.4 The Lorenz equations

It is in the famous works of E. N. Lorenz [21] that a strange attractor appears for the first time. After a drastic simplification of the equations that make possible the study of the behaviour of a fluid in convection, Lorenz was led to study the following system of differential equations:

$$\frac{dx}{dt} = -\sigma x + \sigma y \tag{5}$$

$$\frac{dy}{dt} = rx - y - xz \tag{6}$$

$$\frac{dz}{dt} = -bx + xy \tag{7}$$

where σ, b, r are positive constants that have physical significance. Lorenz made a particular study of the case where $\sigma = 10$, $b = \frac{8}{3}$, $r = 28$.

Recall some simple properties of the Lorenz equations (for more details, see the book of C. Sparrow [29]):

(a) The Lorenz equations have a natural symmetry $(x, y, z) \rightarrow (-x, -y, -z)$ for all values of the parameters.
(b) The z axis is invariant.

(c) They have the origin as a stationary point for all values of the parameters. If $r > 1$, the points C and C′ with coordinates $(\pm\sqrt{b(r-1)}, \pm\sqrt{b(r-1)}, r-1)$ are also stationary points.

Note that if:

$0 < r < 1$: the origin is globally stable;
$1 < r$: the origin is unstable—the linearization of the flow has two negative proper values and one positive;
$1 < r < 27.24\ldots$: the points C and C′ are stable ($\sigma = 10, b = \frac{8}{3}$). The proper values of the flow in a neighbourhood of C and C′ have a negative real part. If $r > 1.346\ldots$, there is a pair of complex conjugate proper values;
$24.27\ldots < r$: the point C and C′ are unstable ($\sigma = 10$, $b = \frac{8}{3}$).

(d) A volume element V is contracted by the flow into an element of volume $V\exp[-(\sigma + b + 1)t]$ at time t (in particular, one cannot find any tori invariant under the action of the flow).
(e) There exists an ellipsoid E such that any trajectory which penetrates E stays caught by E. (To see this, it is enough to consider the Lyapounov function

$$V = rx^2 + \sigma y^2 + \sigma(z - 2r)^2$$

which satisfies

$$\frac{dV}{dt} = -2\sigma(rx^2 + y^2 + bz^2 - 2brz) \tag{8}$$

One can also try other Lyapounov functions.)
(f) There exists a strange attractor; this is a consequence of (d) and (e).

Figure 6.3 shows the Lorenz attractor for $\sigma = 10$, $b = \frac{8}{3}$ and $r = 28$. Note that the trajectories do not intersect despite the appearance of the figure.

Starting from the figure, several 'turbulent' properties can be stated as follows:

(a) The trajectory shown is not periodic.
(b) The figure does not exhibit any transient nature: as long as the numerical integration is continued, the trajectory continues to wind around from one side to the other in an irregular way, with no semblance of periodic or stationary behaviour.
(c) The overall form of the figure does not depend on the choice of initial condition or the method of integration.
(d) The details of the figure *do* depend on the factors mentioned in (c). The series of twists and turns that the trajectory undergoes is entirely sensitive to changes in the initial conditions or of the method of integration. It is consequently impossible to *predict* the detail except for a very short interval of time.

Thus, a finite-dimensional deterministic system can behave in a complicated, turbulent way.

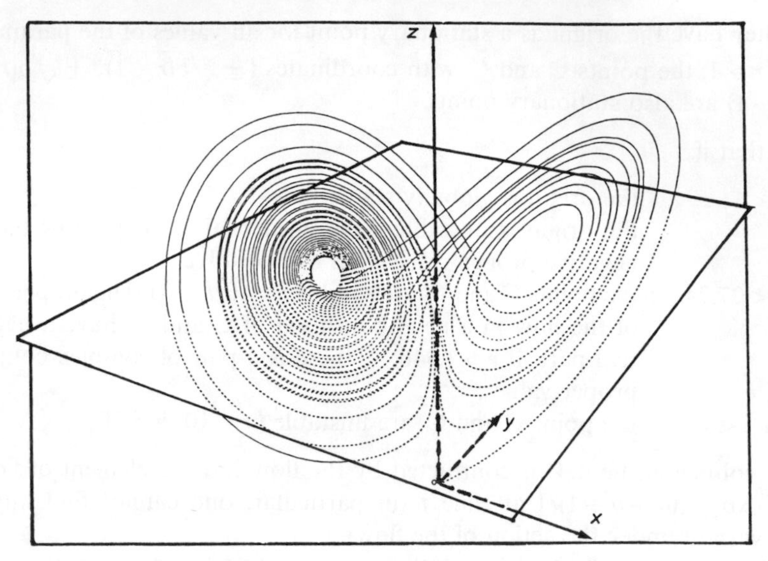

Figure 6.3 The Lorenz attractor, after O. Lanford

We are going to given an upper bound for the Hausdorff dimension of this attractor, using the fundamental theorem and [6]. We write the Lorenz system in the form

$$\frac{dx}{dt} + \sigma(x - y) = 0$$

$$\frac{dy}{dt} + \sigma x + y + xz = 0 \tag{9}$$

$$\frac{dz}{dt} + bz - xy = -b(r + \sigma)$$

making the change of variables $x \to x$, $y \to y$, $z \to z - r - \sigma$. It is then easy to see that from the inequality

$$\frac{1}{2}\frac{d}{dt}|u|^2 + x^2 + y^2 + bz^2 = b(r + \sigma)z$$

$$\leqslant (b-1)z^2 + \frac{b^2}{4(b-1)}(r - \sigma)^2$$

one can deduce that

$$\limsup_{t \to \infty} |u(t)| \leqslant \frac{b(r + \sigma)}{2\sqrt{l(b-1)}}$$

where we have put $u = (x, y, z)$, $l = \min(1, \sigma)$ (we thus again find that with a bounded initial condition, the trajectory remains bounded).

If $S(t)$ is the mapping $u_0 \to u(t)$, its derivative $L(t, u_0)$ is the linear mapping of $\mathbf{R}^3$ into $\mathbf{R}$, $\xi \to L(t, u_0)\xi$ where $L(t, u_0)\xi = U(t)$ is the t-volume of the linearization of (9):

$$U' + \mathscr{A}(u)U = 0, \qquad U(0) = \xi$$

where $\mathscr{A}(u)U = A_1U + A_2U + B(u)U$ and

$$A_1 = \begin{pmatrix} \sigma & 0 & 0 \\ 0 & 1 & 0 \\ 0 & 0 & 2 \end{pmatrix}, \qquad A_2 = \begin{pmatrix} 0 & -\sigma & 0 \\ \sigma & 0 & 0 \\ 0 & 0 & 0 \end{pmatrix}, \qquad B(u) = \begin{pmatrix} 0 & 0 & 0 \\ z & 0 & x \\ -y & -x & 0 \end{pmatrix}.$$

It remains to determine the numbers $\omega_j(L(t, u_0))$, $j = 1, 2, 3$. One can show (see [6]) that

$$\omega_3(L(t, u_0)) \leqslant \exp[-(\sigma + b + 1)t]$$

$$\omega_2(L(t, u_0)) \leqslant \exp(k_2 - \delta)t$$

where $k_2 = -(\sigma + b + 1) + m + b(r + \sigma)/[4\sqrt{l(b-1)}]$, $m = \max(1, b, \sigma)$ and $\delta > 0$ is arbitrarily small. Finally, $\omega_d(L(t, u_0)) < 1$, u_0, if for $d = 1 + s$, we have

$$-s(1 + b + \sigma) + (1 - s)(k_2 - \delta) < 0$$

Since δ is arbitrarily small, one obtains the condition

$$d \leqslant 2 + \frac{k_2}{1 + b + \sigma + k_2}$$

With the usual values $(\frac{8}{3}, 10, 28)$ of the triple (b, σ, r) this 'manual' calculation shows that the Hausdorff dimension d of the Lorenz attractor satisfies $d < 2.538\ldots$. In fact, if numerical simulations are to be trusted, one would expect a value pretty close to 2.06.

6.5 Some point transformations

The study of point transformations is in a fairly advanced state because they can be numerically analysed more conveniently and faster than differential systems (see [13], and particularly its comprehensive bibliography). In what follows, we give a brief account of the Hénon attractor as an example of an attractor with a non-empty interior.

6.5.1 The Hénon attractor

In the wake of the work of E. N. Lorenz and M. Hénon [14], analogous properties were observed in the study of a simple plane point-transformation (for motivation, see the aforementioned article): note that, to start with, Hénon only used a simple H.P. 65! Hénon was thus interested in the point-transformation T

of $\mathbf{R}^2$ into $\mathbf{R}^2$ defined by

$$x_{n+1} = y_n + 1 - ax_n^2$$

$$y_{n+1} = bx_n$$

where a and b are two real parameters such that $a > 0, b \neq 0$. He paid particular attention to the case where $a = 1.4$ and $b = 0.3$. For the mapping T of $\mathbf{R}^2$ into $\mathbf{R}^2$, T^{-1} is defined by

(a) $T^{-1}(x, y) = (b^{-1}y, x - 1 + ab^{-2}y^2)$, (of course, $b \neq 0$)
(b) T has two fixed points whose coordinates are given by

$$x = \frac{-(1-b) \pm (1-b)^2 + 4a}{2a}$$

$$y = bx$$

These points are real if $a > a_0 = (1-b)^2/4$. One fixed point is in the left half-plane, the other in the right half-plane. When $a > a_0$, one of the points is always linearly unstable, while the other is unstable for $a > a_1 = 3(1-b)^2/4$.
(c) The jacobian of T is equal to $-b$; in particular, if $|b| < 1$, areas are contracted under T.
(d) Consider the case where $b = 0.3$.

For $a_0 < a < a_3$, with the initial condition (x_0, y_0), the iterates either go off to infinity or converge to an attractor, which is apparently unique for a given value of a. If, further, $a_0 < a < a_1$, the attractor is a stable fixed point. When

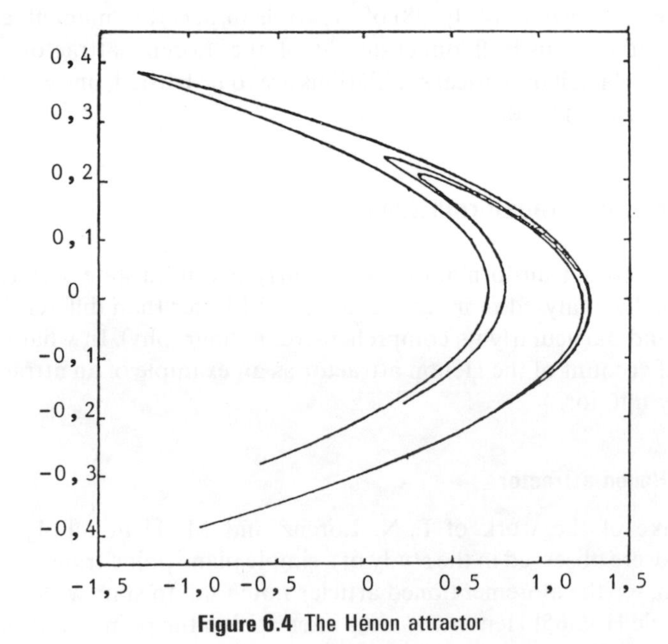

Figure 6.4 The Hénon attractor

a_1 is exceeded, the attractor is still simple and consists of a set of periodic points p. The values of these p increase, after successive 'bifurcations' as a increases, and seem to tend to infinity as a tends to a critical value $a_1 \sim 1.06$ for $b = 0.3$. For $a_2 < a < a_3$, the structure of the attractor becomes complicated and the behaviour of the iterates becomes chaotic (see also the article by J. M. Curry[7]).

Hénon thus observed that, in the case where $a = 1.4$ and $b = 0.3$, the sequence of points obtained by iteration of the transformation T with the given starting-point (x_0, y_0) either tends to infinity or tends to a strange attractor (Figure 6.4).

Considering the initial condition $(x_0, y_0) = (0, 0)$, he established that the fourth iterate is lost within the attractor with the resolution of Figure 6.4; as far as the subsequent points are concerned, they seem to wander erratically over the attractor. If one looks at the local structure of the attractor, successive enlargements of the regions under study reveal the fractal structure of the attractor (see [14] and [26]). Indeed, Hénon has effectively illustrated (see Figure 6.5) a subset U of the plane such that $T(U) \subset U$: it is enough to consider the quadrilateral ABCD whose corners have coordinates

$$x_A = -1.33, \quad y_A = 0.42, \quad x_B = 1.32, \quad y_B = 0.133$$
$$x_C = 1.245, \quad y_C = -0.14, \quad x_D = -1.06, \quad y_D = -0.5$$

(see also [9]).

The dimension of the Hénon attractor is estimated to be about 1.26.

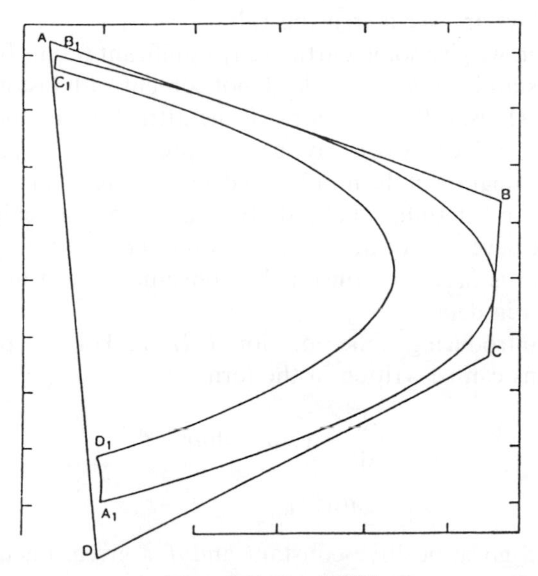

Figure 6.5 *U* and *T*(*U*)

6.5.2 An attractor with non-empty interior

In addition to a certain number of interesting conjectures, Frederickson *et al.* [11] have given several examples of attractors with non-empty interior. Here is one of their examples (for the details, see the aforementioned article).

Consider the point transformation defined by

$$x_{n+1} = 2x_n \bmod 1$$
$$y_{n+1} = y_n + p(x_n)$$

where p is the continuous function defined by

$$\begin{aligned} p(x) &= -4x + 1 && \text{if } 0 \leqslant x < \tfrac{1}{2} \\ &= 4x - 3 && \text{if } \tfrac{1}{2} \leqslant x < 1 \end{aligned}$$

Then, if $\frac{2}{3} \leqslant \lambda < 1$, the preceding transformation has an attractor with a non-empty interior. The fractal dimension of A is thus 2. On the other hand, if $\lambda = \frac{1}{2}$ and $p(x) = \cos 4\pi x$, the authors believe that the preceding transformation has an attractor with an empty interior and of fractal dimension 2.

6.6 The Navier–Stokes equations

For a long time, the study of turbulence seemed to mark time, in spite of the pioneering work of J. Leray [18, 19]. However, there was then quite a change with the popularization of the ideas of L. Landau and E. Lifchitz (see chapter III of [16]), substantially revamped by those of D. Ruelle and F. Takens [28]. It should be said that, in this field, the concept of dimension, without being central to the main results, plays an ever-increasing role.

In this chapter, we give some particularly significant results from the excellent memoir of Constantin *et al.* [6]. Indeed, not content with just giving bounds for the fractal and Hausdorff dimensions of the attractors for the Navier–Stokes equations, they are able to give a physical interpretation: what particularly comes to mind is the estimation of the number of degrees of freedom of a turbulent flow, obtained by more heterodox methods by Landau and Lifchitz [16]. In what follows, we give several typical results with outlines of some proofs; the more interested reader is urged to consult the above-mentioned memoir, where the proofs are given in detail.

We start by introducing some notation. If H is a Hilbert space, the Navier–Stokes equations can be written in the form:

$$\frac{\mathrm{d}u}{\mathrm{d}t} + \nu A u + B(u) = f \tag{10}$$

$$u(0) = u_0 \tag{11}$$

where $\nu > 0$ is a given positive constant and f a given function belonging to $L^\infty((0, \infty), H)$. The operator A is an unbounded, positive, self-adjoint linear operator in H with domain $D(A)$; we denote by (u, v) and $|u|$ the scalar product and the norm of H; furthermore, $D(A)$ is itself a Hilbert space with the scalar product (Au, Av) and the norm $|Au|$. Since the powers A^α, $\alpha\in\mathbf{R}$, can be defined, we may put $V = D(A^{1/2})$ which is a Hilbert space dual to $V' = D(A^{-1/2})$, and V is endowed with the scalar product and norm

$$((u, v)) = (A^{1/2}u, A^{1/2}v), \qquad \|u\| = |A^{1/2}u|$$

Recall that A possesses an orthonormal family of proper vectors ω_j, $j \geqslant 1$, and that this family is complete in H:

$$A\omega_j = \lambda_j\omega_j, \qquad j \geqslant 1, \qquad 0 < \lambda_1 \leqslant \lambda_2, \ldots, \lambda_j \to +\infty \qquad \text{as } j \to \infty$$

For B, we have $B(u) = B(u, u)$ where $B(\cdot,\cdot)$ is a bilinear operator with good continuity properties.

More mundanely, $H, V, D(A), A, B, \ldots$ will be defined in the following way:

$$H = \{v\in L^2(\Omega)^3, \operatorname{div} v = 0, v\cdot\mathbf{v} = 0 \text{ on } \partial\Omega\}$$

where $\mathbf{v}$ is the outer normal to $\partial\Omega$(Ω being a bounded domain of $\mathbf{R}^3$):

$$V = \{v\in H_0^1(\Omega)^3, \operatorname{div} v = 0\}$$

$$D(A) = H_0^1(\Omega)^3 \cap H^2(\Omega)^3$$

$$Au = -P\Delta u$$

$$B(u, v) = P(u\cdot\nabla)v)$$

where P is the orthogonal projection of $L^2(\Omega)^3$ onto H.

Further, we have

$$(u, v) = \int_\Omega u(x)v(x)\,\mathrm{d}x$$

$$((u, v)) = \sum_{i,j=1}^{3} \int_\Omega \frac{\partial u_i}{\partial x_j}(x)\frac{\partial v_i}{\partial x_j}\,\mathrm{d}x$$

It is then easy to see that

$$\forall v\in V, \qquad \|v\|^2 = ((v, v)) = \int_\Omega |\operatorname{rot} v|^2\,\mathrm{d}x$$

(The quantity $\frac{1}{2}\|v\|^2$ is the attractor of the field v.)

Definition Let $u_0\in V$, $f\in L^\infty((0, \infty), H)$. A *strong solution* of the problem of (10) and (11) with the initial condition defined on an interval $[0, T]$, $T > 0$ is a function $u\in L^\infty((0, T); V)\cap L^2((0, T); D(A))$ which satisfies (10) on $[0, T]$ and (11).

It is known that a strong solution exists and is unique on an interval $[0, T_1]$ where T_1 is of the form

$$T_1 = T_1(\|u_0\|) = \frac{K_1}{(1 + \|u_0\|^2)^2}$$

where K_1 only depends on $|f|_{L^\infty((0,\infty);\,H)}$, ν and Ω (see the references in [6]).

For given $t > 0$, we denote by $S(t)$ the mapping $V \ni u_0 \to u(t) \in V$ that assigns to u_0 the value at time t of the solution of the Navier–Stokes equations (10) and (11), when this mapping is defined, i.e. when the solution satisfies (10) and (11); thus

$$u|_{[0,t]} \in L^\infty((0,t);\ V)$$

Thus $S(t)u_0$ is well defined for every u_0 such that $\|u_0\| \leqslant R$ and for each $t, 0 < t \leqslant T_1(R)$.

Let u be a solution of the Navier–Stokes equations on $(0,\infty)$ which is uniformly bounded in V, i.e. such that $\|u\|_{L^\infty((0,\infty);\,V)} \leqslant K_2 < \infty$, where K_2 only depends on $|f|_{L^\infty((0,\infty);\,V)}$, ν and Ω. Consider the set X defined by

$$X = \bigcap_{t>0} \overline{\{u(s), s \geqslant t\}}$$

where the closure is taken in H. It is shown by C. Foias and R. Temam in [10] that X possesses the following properties:

(*) X is bounded in V, compact in H.
(**) $S(t)u_0$ is defined for each $u_0 \in X$ and for every $t > 0$; furthermore, $S(t)X = X, \forall t > 0$.

Definition A set $X \subset H$ which possesses properties (*) and (**) is called an *invariant functional set* (or an attractor). We are going to give some theorems that provide a bound for the fractal and Hausdorff dimensions of an attractor associated with a three-dimensional flow. We will suppose that f is *independent of t*.

To start with, here is a theorem proved in [6] by Constantin *et al.*

Theorem 3 Let X be a set that is a functional invariant set for the Navier–Stokes operators and is bounded in V. Let $\mathscr{A}(u)$ be the operator from $D(A)$ into H defined by

$$\mathscr{A}(u)U = \nu AU + B(u, U) + B(U, u)$$

Letting Ω denote a rank-m projection, put

$$-g_m = \limsup_{t\to\infty} \left\{ \inf \frac{1}{t} \int_0^t \inf_{\text{rank } Q = m} (\mathrm{Tr}(\mathscr{A}(u(s)) o Q)\, \mathrm{d}S \right\}$$

Then, if there exists m such that $q_m > 0$, we have

$$d_H(X) \leqslant m \tag{12}$$

$$d_F(X) < \infty \tag{13}$$

More precisely,

$$d_F(X) \leqslant m\left(1 + \max_{1 \leqslant l \leqslant m} \frac{-q_l}{q_m}\right)$$

Remark Of course, other authors have considered the Hausdorff dimension of the attractors for the Navier–Stokes equations: O. A. Ladyshenskaya [15], A. V. Babin and M. I. Vishnik [1, 2, 3], D. Ruelle [27], E. Lieb [20], for example. Note, however, that while the result of Lieb is somewhat better than that of the above theorem, this author does not concern himself with the Hausdorff dimension.

Sketch of the proof It is enough to apply the fundamental theorem mentioned above. We shall suppose that the delicate question of differentiability in the sense of condition (2) of Section 6.3 has been settled. Let, therefore, $L(t, u)$ be the derivative of $S(t)$ at u. The numbers $\omega_j(L(t, u))$, $\bar{\omega}_j(t) = \sup_{u \in X} \omega_j(L(t, u))$ are well defined, just like the numbers Π_j, Λ_j, μ_j. It is necessary to give an estimate of $\omega_m(L(t, u_0))$ and of $\limsup_{t \to \infty} \sup_{u_0 \in X} \{\omega_m(L(t, u_0))\}^{1/t}$. Wc have

$$\omega_m(L(t, u_0)) = \sup_{\substack{|\xi_j| \leqslant 1 \\ 1 \leqslant j \leqslant m}} |U_1(t) \wedge \cdots \wedge U_m(t)|$$

where $j = 1, \ldots, m$ and U_j denotes the solution of

$$\frac{dv}{dt} + \nu AU + B(u, U) + B(U, u) = 0$$

$$U(0) = \xi$$

(u being a solution of (10) and (11). A little calculation shows that

$$\frac{d}{dt}|U_1 \wedge \cdots \wedge U_m|^2 + \mathrm{Tr}(\mathscr{A}(u) o Q_m)|U_1 \wedge \cdots \wedge U_m|^2 = 0$$

where Q_m is the orthogonal projection in H on the subspace generated by $U_1, \ldots, U_m$. It follows that

$$|U_1 \wedge \cdots \wedge U_m| = |\xi_1 \wedge \cdots \wedge \xi_m| \exp\left(-\int_0^t \mathrm{Tr}(\mathscr{A}(u) o Q_m)\, ds\right)$$

so that

$$\omega_m(L(t, u_0)) \leqslant \exp\left(-\int_0^t \mathrm{Tr}(\mathscr{A}(u) o Q_m)\, ds\right)$$

Finally, we have

$$\bar{\omega}_m(t) = \sup_{u_0 \in X} \omega_n(L(t, u_0)) \leqslant \sup_{u_0 \in X} \exp\left(-\int_0^t \inf_{\mathrm{rank}\ Q = m} \mathrm{Tr}(\mathscr{A}(u) o Q)\, ds\right)$$

so that

$$\Pi_m = \lim_{t \to \infty} \bar{\omega}_m(t)^{1/t} \leqslant \exp(-q_m)$$

It is the above theorem that enables Constantin, Foias and Temam to give an upper bound for the number of degrees of freedom in terms of the Kolmogorov length.

At this point, we introduce some more notation. Let X be a bounded attractor of V such that

$$\sup_{v\in X}\int_{\Omega}|\nabla v|^2\,\mathrm{d}x<\infty$$

It is known that if $f\in H^1(\Omega)^3\cap H$, then $X\subset H^3(\Omega)^3$: this is a result of C. Guillopé [12]; we also know, by the Sobolev embedding theorems in several dimensions, that $H^3(\Omega)\subset\mathscr{C}^1(\Omega)$. Thus, $\sup_{v\in X}\sup_{x\in\Omega}|\nabla v(x)|<\infty$ and we can define, following Landau and Lifchitz [16], the rate of dissipation of energy per unit mass and time by

$$\varepsilon=\nu\limsup_{t\to\infty}\left\{\sup_{u_0\in X}\frac{1}{t}\int_0^t\sup_{x\in\Omega}|\nabla u(\tau,x)|^2\,\mathrm{d}\tau\right\}$$

where $u(t,\cdot)=S(t)u_0$. This is the time average of the maximum rate of dissipation.

The *Kolmogorov length* is defined by

$$l_d=\left(\frac{\nu^3}{\varepsilon}\right)^{1/4}$$

and the typical length by $l_0=\lambda_1^{-1/2}$.

Theorem 4 If X is a bounded invariant functional set in V, we have $d_F(X)\leqslant c(l_0/l_d)^3$ where c is a universal constant.

Sketch of the proof We need to give an estimate for q_m. To this end, let Q be a rank-m orthogonal projection in H. Let $(\Phi_j)_{j\in\mathrm{N}}$ be an orthonormal basis in H such that $\Phi_1,\ldots,\Phi_m$ is a basis for QH. Then

$$\begin{aligned}\mathrm{Tr}\mathscr{A}(u)\mathrm{o}Q&=\sum_{j=1}^{m}(\mathscr{A}(u)\Phi_j,\Phi_j)\\&=\sum\{\nu(A\Phi_j,\Phi_j)+B(\Phi_j,u)+(B(u,\Phi_j),\Phi_j)\\&=\nu\,\mathrm{Tr}\,A\mathrm{o}Q+\sum_{j=1}^{m}(B(\Phi_j,u),\Phi_j)\end{aligned}$$

Now

$$\left|\sum_{j=1}^{m}(B(\Phi_j,u),\Phi_j)\right|\leqslant m|\nabla u|_{L^\infty(\Omega)}$$

We still have to find a lower bound for $\mathrm{Tr}(A\mathrm{o}Q)$; now, it is shown in [6] that $\mathrm{Tr}\,A\mathrm{o}Q\geqslant\lambda_1+\cdots+\lambda_m$. By a result of G. Métivier, there exists a universal constant c_1' such that $\lambda_m\geqslant c_1'm^{2/3}\lambda_1$, for all $m\geqslant 1$.

It follows from this that

$$\nu\,\mathrm{Tr}\,A\mathrm{o}Q\geqslant\nu\lambda_1c_1'\sum_{j=1}^{m}J^{2/3}\geqslant\nu\lambda_1c_2'm^{5/3}$$

and finally,

$$-\operatorname{Tr}(\mathscr{A}(u) \mathrm{o} Q) \leqslant -\nu\lambda_1 c_2' m^{5/3} + m|\nabla u|_{L^\infty(\Omega)}$$

If $u = S(t)u_0$, with $u_0 \in X$, we have

$$\begin{aligned} -\frac{1}{t}\int_0^t \operatorname{Tr}(\mathscr{A}(u)\mathrm{o}Q)\,\mathrm{d}s &\leqslant -\nu\lambda_1 c_2' m^{5/3} + \frac{m}{t}\int_0^t |\nabla u|_{L^\infty(\Omega)}\,\mathrm{d}\tau \\ &\leqslant -\nu\lambda_1 c_2' m^{5/3} + m\left(\frac{1}{t}\int_0^t |\nabla u|^2_{L^\infty(\Omega)}\,\mathrm{d}\tau\right)^{1/2} \\ &\leqslant -\nu\lambda_1 c_2' m^{5/3} + m\left(\frac{\varepsilon}{\nu}\right)^{1/2} \end{aligned}$$

Thus,

$$q_m \geqslant \nu\lambda_1 c_2' m^{5/3} - m\left(\frac{\varepsilon}{\nu}\right)^{1/2}$$

and

$$q_m > 0 \text{ as long as } m^{2/3} > \frac{1}{c_2'\lambda_1}\left(\frac{\varepsilon}{\nu^3}\right)^{1/2} = \frac{1}{c_2'}\left(\frac{l_0}{l_d}\right)^2$$

Let m be the least integer such that

$$m^{2/3} \geqslant \frac{2}{c_2'}\left(\frac{l_0}{l_d}\right)^2$$

so that

$$m - 1 < \left(\frac{2}{c_2'}\right)^{3/2}\left(\frac{l_0}{l_d}\right)^3 \leqslant m$$

Put

$$\Theta = \frac{1}{c_2'}\left(\frac{l_0}{l_d}\right)^2$$

By a similar argument we obtain

$$-\frac{q_{j'}}{q_m} \leqslant -\frac{j^{5/3} + j\Theta}{m^{5/3} - m\Theta} \leqslant \frac{l_j\Theta}{m^{5/3}} \leqslant 1, \qquad j = 1, \ldots, m-1$$

Finally,

$$d_F(X) \leqslant 2m \leqslant (2 + 2\Theta)^{3/2} \leqslant \left[2 + \left(\frac{2}{c_2'}\right)\right]^{3/2}\left(\frac{l_0}{l_d}\right)^3$$

We now give an estimate *in terms of the Reynolds number*. Let

$$\mathrm{Re} = \frac{1}{\nu\lambda_1^{1/2}} \limsup_{t\to\infty} \sup_{u_0 \in X}\left\{\left(\frac{1}{t}\int_0^1 |S(\tau)u_0|^2_{L^\infty(\Omega)}\,\mathrm{d}\tau\right)^{1/2}\right\}$$

the Reynolds number associated with the time average of the maximum absolute value of the speed on the attractor.

Theorem 5 If X is a bounded attractor in V, then

$$d_F(X) \leqslant c_1 \mathrm{Re}^3$$

where c_1 is a universal constant.

Sketch of the proof With the above notation, we have

$$-\mathrm{Tr}(\mathscr{A}(u)\mathrm{o}Q) = -\nu \sum_{j=1}^{m} \|\Phi_j\|^2 - \sum_{j=1}^{m} (B(\Phi_j, u), \Phi_j)$$

A little calculation yield the following bound:

$$\begin{aligned}-\mathrm{Tr}(\mathscr{A}(u)\mathrm{o}Q) &\leqslant -\frac{\nu}{2}\sum_{j=1}^{m} \|\Phi_j\|^2 + \frac{m}{2}|u|^2_{L^\infty(\Omega)} \\ &\leqslant -\frac{\nu}{2}\mathrm{Tr}(\mathscr{A}\mathrm{o}Q) + m|u|^2_{L^\infty(\Omega)} \\ &\leqslant -\frac{\gamma\lambda_1}{2}c_2' m^{5/3} + \frac{m}{2}|u|^2_{L^\infty(\Omega)}\end{aligned}$$

whence

$$\limsup_{t\to\infty} \sup_{u_0\in X}\left\{-\frac{1}{t}\int_0^t \inf_{\mathrm{rank}\ Q=m} \mathrm{Tr}(\mathscr{A}(u)\mathrm{o}Q\,\mathrm{ds}\right\}$$

$$\leqslant -\frac{\nu\lambda_1}{2}c_2' m^{5/3} + \frac{\nu\lambda_1 m}{2}\mathrm{Re}^2$$

We finally deduce from this that

$$q_m \geqslant \frac{\nu\lambda_1}{2}c_2' m^{5/3} - \frac{\nu\lambda_1 m}{2}\mathrm{Re}^2$$

Let m be the smallest integer such that $m^{2/3} \geqslant (2/c_2')\,\mathrm{Re}^2$. Arguing as above and using the fundamental theorem, we obtain

$$d_F(X) \leqslant 2m \leqslant c_1 \mathrm{Re}^3$$

We give an estimate using another Reynolds number. Let

$$G = \frac{|f|_\infty}{v^2\lambda_1^{3/4}}$$

be the Grashoff number. Let

$$\overline{\mathrm{Re}} = \frac{1}{\nu\lambda_1^{1/2}} \sup_{u\in\Omega}\sup_{x\in\Omega}|u(x)|$$

be the Reynolds number associated with the maximum absolute value of the velocity vector on the attractor.

Theorem 6 If X is a bounded attractor in V, then

$$d_F(X) \leqslant c_2 G^3 (1 + \overline{\text{Re}}^{3/2})$$

Concluding remark In their article, Constantin, Foias and Terman [6] give a fourth bound for the fractal dimension of an attractor using another physical quantity. Remarkably enough, the authors are able to achieve an improvement to most of the preceding theorems with the help of strengthened Lieb and Thirring inequalities!

References

1. Babin A.V. and Vishnik M. T. (1982) Attractors of quasilinear parabolic equations, *Doklady Ak. N. SSSR*, **254**(4), 780–784.
2. Babin A. V. and Vishnik M. I. (1982) Existence of attractors and estimate for the dimension of attractors in quasi-linear parabolic equations and Navier–Stokes systems, *Usp. Math. Nauk.*, **37**(3).
3. Babin A. V. and Vishnik M. I. (1983) Regular attractors of semigroups and evolution equations, *J. Math. Pures Applic.*, **62**, 441–491.
4. Berge P., Pomeau Y. and Vidal Ch. (1984) *L'Ordre dans le Chaos: vers une Approche Déterministe de la Turbulence*, Hermann, Paris.
5. Constantin P. and Foias C. Global Lyapounov Exponents—Kaplan Yorke formulas and the dimension of the attractors for 2-D Navier–Stokes equations, *Comm. Pure Applic. Math.*
6. Constantin P., Foias C. and Temam R. (1985) Attractors representing turbulent flows, Memoirs of the American Mathematical Society **314**.
7. Curry J. H. (1979) On the Hénon transformation, *Comm. Math. Phys.*, **68**, 129–140.
8. Douady A. and Oesterle J. (1980) Dimension de Hausdorff des attracteurs. CRAS, **290**, Series A, 1135–1138.
9. Feit S. D. (1978) Characteristic exponents and strange attractors, *Comm. Math. Phys.*, **61**, 249–260.
10. Foias C. and Temam R. (1979) Some analytic and geometric properties of the solutions of the Navier–Stokes equations, *J. Math. Pure Applic.*, **58**, 339–368.
11. Frederickson P., Kaplan J. L., York E. D. and Yorke J. A. (1983) The Lyapounov dimension of strange attractors, *J. Diff. Equat.*, **49**, 185–207.
12. Guillopé C. (1982) Comportement a l'infini des équations de Navier–Stokes et propriété des ensembles fonctionnels invariants (ou attracteurs), *Ann. Inst. Fourier (Grenoble)*, **1982**, 32–33.
13. Gumowski I. and Mira C. (1980) *Dynamique Chaotique: Transformations Ponctuelles—Transition: Ordre–Désordre*, Cepadues Editions, Toulouse.
14. Hénon M. (1976) A two-dimensional mapping with a strange-attractor, *Comm. Math. Phys.*, **50**, 69–76.
15. Ladyzhenskaya O. A. On the finiteness of the dimension of bounded sets for the Navier–Stokes equations and other related dissipative systems, in *The Boundary*

Value Problems of Mathematical Physics and Related Questions in Functional Analysis, Vol. 14, Seminar of the Steklov Institute, Leningrad.
16. Landau L. and Lifchitz E. (1971) *Fluid Mechanics*, Editions MIR, Moscow.
17. Ledrappier F. (1981) Some relations between dimension and Lyapounov exponents, *Comm. Math. Phys.*, **81**, 229–238.
18. Leray J. (1933) Etude de diverses équations intégrales non linéaires et de quelques problèmes que pose l'hydrodynamique, *J. Math. Phys. Pures Applic.*, **12**, 1–82.
19. Leray J. (1934) Essai sur les mouvements plans d'un liquide visqueux que limitent des parois, *J. Math. Pures Applic.*, **13**, 331–418.
20. Lieb E. (1984) On characteristics exponents in turbulence, *Comm. Math. Phys.*, **92**, 473–480.
21. Lorenz E. N. (1983) Deterministic non-periodic flow, *J. Atmos. Sci.*, **20**, 130–141.
22. Mandelbrot B. (1977) *Fractals: Form, Chance and Dimension*, Freeman; San Francisco.
23. Mane R. (1980) On the dimension of the compact invariant sets of certain non-linear maps, in *Dynamical Systems and Turbulene*, Warwick, Lecture Notes 898.
24. Marsden J. E. and McCracken M. (1976) *The Hopf Bifurcation and Its Applications*, Applied Mathematical Sciences, Springer-Verlag, New York.
25. Milnor J. (1985) on the concept of attractor, *Comm. Math. Phys.*, **99**, 177–195.
26. Ruelle D. (1979–1980) Strange attractors, *Math. Intell.*, **2**, 126–137.
27. Ruelle D. (1984) Characteristic exponents for a viscous fluid subjected to time-dependent flows, *Comm. Math. Phys.*, **98**, 285–300.
28. Ruelle D. and Takens F. (1971) On the nature of turbulence, *Comm. Math. Phys.*, **20**, 167–192, 343–344.
29. Sparrow C. (1982) *The Lorenz Equations: Bifurcations, Chaos and Strange Attractors*, Springer-Verlag, New York.
30. Weber M. (1985) Processus stochastiques et Procédés de recouvrement, Séminaire Interdisciplinaire Hausdorff, Laboratoire TMIB, Université Paris VII.
31. Young L. S. (1981) Capacity of attractors, *Ergodic Th. and Dynam. Syst.*, **1**, 381–383.
32. Young L. S. (1982) Dimension entropy and Lyapounov exponents, *Ergod. Th. and Dynam. Syst.*, **2**, 109–124.

Final note The author also cites the following article, which he only discovered after the paper was written.

Campbell D. and Rose H. (1982) Order in chaos, in *Proc. of the International Conference on Order in Chaos*, held at the Centre for Non-linear Studies, Los Alamos, New Mexico, North-Holland Publishers.

Chapter 7 Construction of fractals and dimension problems

F. M. Dekking

Department of Mathematics, University of Delft

We are going to discuss a mathematical description of a broad class of fractal objects. This class includes all the classical examples, such as, for instance, the Peano and von Koch curves, etc., and the totally disconnected sets of Cantor, Sièrpinski, Menger, etc. This point of view enables one to give a precise description of such objects, to classify them and to determine their Hausdorff dimension in many cases. Lastly, this theory leads, in a very simple way, to mechanical realizations of (approximations to) fractals.

In the first section, we present an introduction to this description. In the second, we discuss several generalizations of a well-known property associated with the notion of fractal: self-similarity. The third section is given over to the problems of calculating the Hausdorff dimension.

7.1 A mathematical description of fractals

Consider the construction of the surve conceived by Peano and published in 1890. The starting-point is a straight line segment K_0 that is replaced by a polygonal line K_1 (see Figure 7.1). To obtain K_2, once again replaces each of the nine segments of K_1 by a polygonal line similar to K_1. In order to describe this construction , we code the four possible directions in the curves K_n by means of the letters e, n, w, s (see Figure 7.2). The code for K_0 is then the letter e, and that of K_1 is the word W_1 formed by the nine letters *eneswsene*. Having no desire to write out the 81-letter word W_2 coding K_3 (the curve obtained from K_2 by again replacing each of its segments by a polygonal line similar to K_1), we give here a list of words that replace letters to allow one to obtain W_2 from W_1, in general, W_{n+1} from W_n:

$e \rightarrow eneswsene$
$n \rightarrow nwnesenwn$
$w \rightarrow wswnenwsw$
$s \rightarrow seswnwses$

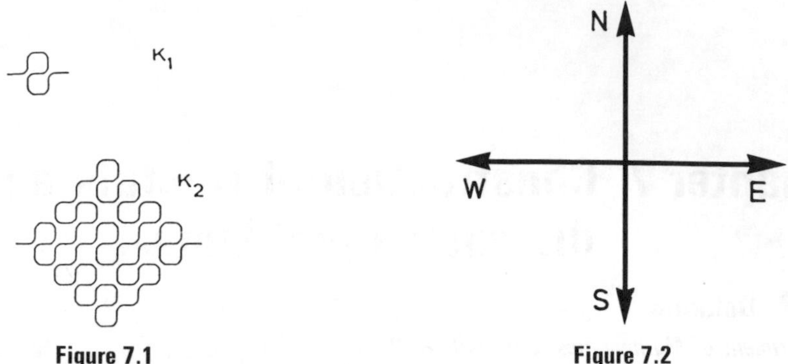

Figure 7.1

Figure 7.2

Mathematically, these replacements describe a morphism (called substitution) of the set $S^* = \cup_{k \geqslant 0} S^k$ of all words into itself, where $S = \{e, n, w, s\}$ is the set of letters. The above remarks tell us that K_n is coded by the word $W_n = \theta^n(e)$. Here, the powers θ^n are the iterates of θ defined by $\theta^1 = \theta$, $\theta^{n+1}(W) = \theta(\theta^n(W))$ if $W \in S^*$. One uses the word morphism because θ must respect the operation of concatenation of words: $\theta(WW') = \theta(W)\theta(W')$.

We cannot stop at this because we do not yet have a geometrical description of the curve. In greater generality, consider an alphabet S together with a mapping $f: S^* \to \mathbf{R}^d$ (d-dimensional euclidean space) satisfying $f(WW') = f(W) + f(W')$. The mapping f allows one to associate with each word W a *point* in $\mathbf{R}^d$. To obtain *curves*, or more general objects, one considers a mapping $K[\cdot]: S^* \to (\mathbf{R}^d)$, the set of compact subsets of $\mathbf{R}^d$ possessing the property

$$(*) K[WW'] = K[W] \cup (K[W'] + f(W)), \qquad W, W' \in S^*$$

We have here used the following notation: for $A \subset (\mathbf{R}^d)$ and $x \in \mathbf{R}^d$, $A + x = \{a + x : a \in A\}$. The canonical choice for $K[\cdot]$ is given by

$$K[s] = \{\alpha f(s) : 0 \leqslant \alpha \leqslant 1\}, \qquad s \in S$$

(Note that, just like $f(\cdot)$, $K[\cdot]$ is determined by its values on the letters.) One easily verifies that in this case $K[W]$ is the polygonal line passing through the origin, $f(m_1), f(m_1 m_2), \ldots, f(m_1 m_2 \cdots m_p)$ for each word $W = m_1 m_2 \cdots m_p$.

In the Peano curve example, one chooses $f(e) = \begin{pmatrix} 1 \\ 0 \end{pmatrix}$, $f(n) = \begin{pmatrix} 0 \\ 1 \end{pmatrix}$, $f(w) = \begin{pmatrix} -1 \\ 0 \end{pmatrix}$ and $f(s) = \begin{pmatrix} 0 \\ -1 \end{pmatrix}$. With the canonical choice for $K[\cdot]$, K_n is equal to $K[\theta^n(e)]$, to within a contraction. For this, let L be the linear mapping given by $L(x) = 3x$ for $x \in \mathbf{R}^2$. We then finally have $K_n = L^{-n} K[\theta^n(e)]$, where the K_n are the Peano curves (Figure 7.3).

The fact that the relation $f(\theta(t)) = L(f(t))$ holds for each $t \in \{e, n, w, s\}$ corresponds to the property that each vertex of K_n is also a vertex of K_{n+1}, which gives some 'compatibility' between the K_n. In the general case, we say that a substitution $\theta: S^* \to S^*$ admits a *representation* in $\mathbf{R}^d$ if there exists a homomorph-

Figure 7.3

isms $f:S^* \to \mathbf{R}^d$ and a linear mapping $L:\mathbf{R}^d \to \mathbf{R}^d$ such that

$$f(\theta(s)) = L(f(s)), \qquad s \in S$$

L is said to be *expansive* if all its proper values are of modulus greater than one.

Theorem 1 ([2]) Let $\theta:S^* \to S^*$ be a substitution with an expansive representation L and $K[\cdot]:S^* \to (\mathbf{R}^d)$ a mapping with the property (*), such that $K[s] \neq \phi$ for $s \in S$. Then there exists, for each word W, a non-empty compact set $K_\theta(W)$ such that

$$L^{-n}K[\theta^n(W)] \to K_\theta(W), \qquad \text{as } n \to \infty$$

in the Hausdorff metric. The set $K_\theta(W)$ does not depend on the choice of $K[\cdot]$ and is a Jordan curve, i.e. the continuous image of an interval.

Note Recall that two compact set A and B are less than the distance $\varepsilon > 0$ apart in the Hausdorff metric if $A \subset B^\varepsilon$ and $B \subset A^\varepsilon$, where $A^\varepsilon = \{x: \|x - a\| < \varepsilon, a \in A\}$.

Example 1 To obtain the von Koch curve, take $S = \{a, b, c, d, e, f\}$, θ defined by $\theta(a) = abfa$, $\theta(b) = bcab$, etc. (extended by symmetry), $f:S^* \to \mathbf{R}^2$ defined by $f(a) = \begin{pmatrix} 1 \\ 0 \end{pmatrix}$, $\quad f(b) = \begin{pmatrix} \frac{1}{2} \\ \frac{1}{2}\sqrt{3} \end{pmatrix}$, etc. Then $f\theta = Lf$ where L is again multiplication by 3. The theorem may be applied. Taking $W = a$, one obtains the von

Koch curve. To illustrate the fact that the limit does not depend on the choice of $K[\cdot]$, we have taken $K[s] = *$ for $s \in S$ in Figure 7.4.

Under the conditions of Theorem 1, one only obtains fractal curves. In order to obtain the totally disconnected sets, we introduce the following notion. A letter s for which $K[s] = \phi$, the empty set, is called *virtual*. Let Q be the set of virtual letters. A letter s with the property that $\theta^k(s) \notin Q^*$ for almost every positive integer k is *essential*. We denote by E the set of essential letters.

Example 2 Let $S = \{a, b, c, \bar{a}, \bar{b}, \bar{c}\}$ and $Q = \{\bar{a}, \bar{b}, \bar{c}\}$, which means that $K[\bar{a}] = K[\bar{b}] = K[\bar{c}] = \phi$, but $K[a]$, $K[b]$ and $K[c]$ are all non-empty. Let θ be defined

Figure 7.4

by $\theta(a)=\bar{a}b\bar{a}$, $\theta(\bar{a})=a\bar{b}a$, $\theta(b)=\bar{b}c\bar{b}$, $\theta(\bar{b})=b\bar{c}b$, $\theta(c)=\bar{c}a\bar{c}$, $\theta(\bar{c})=c\bar{a}c$. All letters are then essential.

Example 3 Let $S=\{a,\bar{a}\}$, $K[a]=[0,1]$ and $K[\bar{a}]=\phi$. Then $Q=\{\bar{a}\}$. Let θ be given by $\theta(a)=a\bar{a}a$ and $\theta(\bar{a})=\bar{a}\bar{a}\bar{a}$. This substitution admits a representation in $\mathbf{R}$: take $f(a)=f(\bar{a})=1$ and again let L be multiplication by 3. In this example $E=\{a\}=S/Q$. It is not hard to convince oneself that

$$\lim_{n\to\infty} L^{-n}K[\theta^n(a)]=C$$

where C is the Cantor ternary set.

The substitution θ is said to be *Q-stable* if each letter s is either essential or satisfies $\theta^k(s)\in Q^*$ for almost every positive integer k.

Theorem 2 ([2]) Let $\theta:S^*\to S^*$ be a substitution with expansive representation L, and $K[\cdot]:S^*\to\mathscr{K}(\mathbf{R}^d)$ a mapping with the property (*). Let Q be the set of virtual letters. Suppose that θ is Q-stable. Then for each word W with $K[W]\neq\phi$ there exists a non-empty compact set $K_\theta(W)$ such that

$$L^{-n}K[\theta^n(W)]\to K_\theta(W), \qquad \text{as } n\to\infty$$

in the Hausdorff metric.

This theorem is applicable in the case of Examples 2 and 3. The substitution in Example 2 admits a representation L in $\mathbf{R}^2$ (multiplication by 3, followed

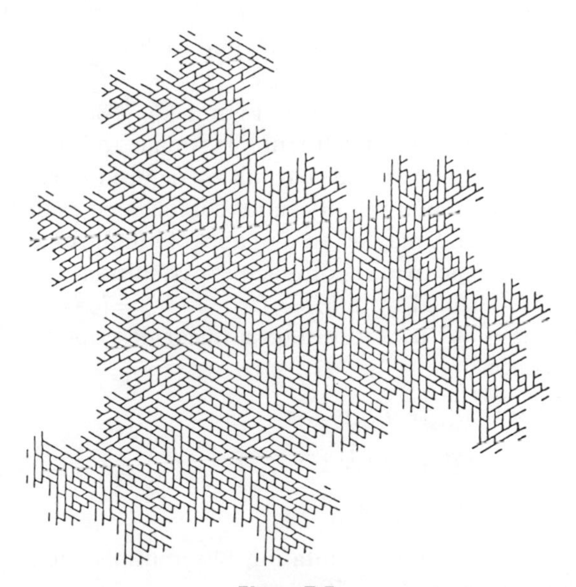

Figure 7.5

by rotation through $\pi/6$), taking $f(a) = f(\bar{a}) = \begin{pmatrix} 1 \\ 0 \end{pmatrix}$, $f(b) = f(\bar{b}) = \begin{pmatrix} -\frac{1}{2} \\ \frac{1}{2}\sqrt{3} \end{pmatrix}$ and $f(c) = f(\bar{c}) = \begin{pmatrix} -\frac{1}{2} \\ \frac{1}{2}\sqrt{3} \end{pmatrix}$.

Figure 7.5 shows

$$L^{-7}K[\theta^7(abc)]$$

where $K[\cdot]$ is the canonical choice.

There is an important difference between the two examples 2 and 3. The interesting paving that is obtained with the $L^{-n}K[\theta^n(abc)]$ of Example 2 is a *finite* property of the construction due to an interweaving of virtual and non-virtual letters in θ. It is in fact proved in [4] that each fractal $K_\theta(W)$ in Theorem 2 can be obtained by means of an S' and θ', with Q' and E' respectively the virtual and essential letters in S', such that $E' = S'/Q'$ and $\theta'(Q'^*) \subset Q'^*$. Thus the fractal in Example 2 is none other than the 'terdragon' of Davis and Knuth or the 'fudge flake' of Mandelbrot [9].

7.2 Self-similarity properties

The classical examples of fractal sets often possess a self-similarity property: the fractal F may be decomposed into a finite number of subsets $F_1, \ldots, F_p$ all of which are similar to F and whose pairwise intersections are either empty or negligible (of lower dimension, for example). An initial generalization of this property suggests itself naturaly in the context of the formalism of Section 7.1. The fractals $F^{(1)}, \ldots, F^{(m)}$ are said to have the *multi-type self-similarity* property if each $F^{(i)}$ may be decomposed into subsets $F_j^{(i)}$, $1 \leqslant j \leqslant P_i$, such that $F_j^{(i)} \cap F_{j'}^{(i)}$ is empty (or else negligible) for $j \neq j'$ and such that each $F_j^{(i)}$ is similar to one of the $F^{(1)}, \ldots, F^{(m)}$.

A natural example of this is a tree (see Figure 7.6). There are two types here: the trunk t and the branches b. Each trunk is similar to itself and each branch is formed from three branches and a trunk of reduced size. We are going to give a precise description of a slightly more complicated example. Let $S\{b_\alpha, t_\alpha, V_\alpha{:}0 \leqslant \alpha < 2\pi\}$. Although S may be an alphabet of infinite cardinality, Theorem 2 is still valid under some (weak) compactness conditions on f and $K[\cdot]$. The homomorphism f is defined by $f(b_\alpha) = f(t_\alpha) = f(v_\alpha) = (\cos\alpha, \sin\alpha)$. The v_α are the virtual letters that one uses to obtain a linear description of the tree. The substitution is defined by

$$\theta(t_\alpha) = t_\alpha b_\alpha t_\alpha$$

$$\theta(b_\alpha) = t_\alpha t_\alpha b_{\alpha+\beta} v_{\alpha+\beta-\pi} b_{\alpha-\beta} v_{\alpha-\beta-\pi} b_\alpha$$

$$\theta(v_\alpha) = v_\alpha v_\alpha v_\alpha$$

where β is a parameter. The representation L is multiplication by 3. Figure 7.7. shows $L^{-5}K[\theta^5(t_{\pi/2})]$, with $\beta = \pi/10$.

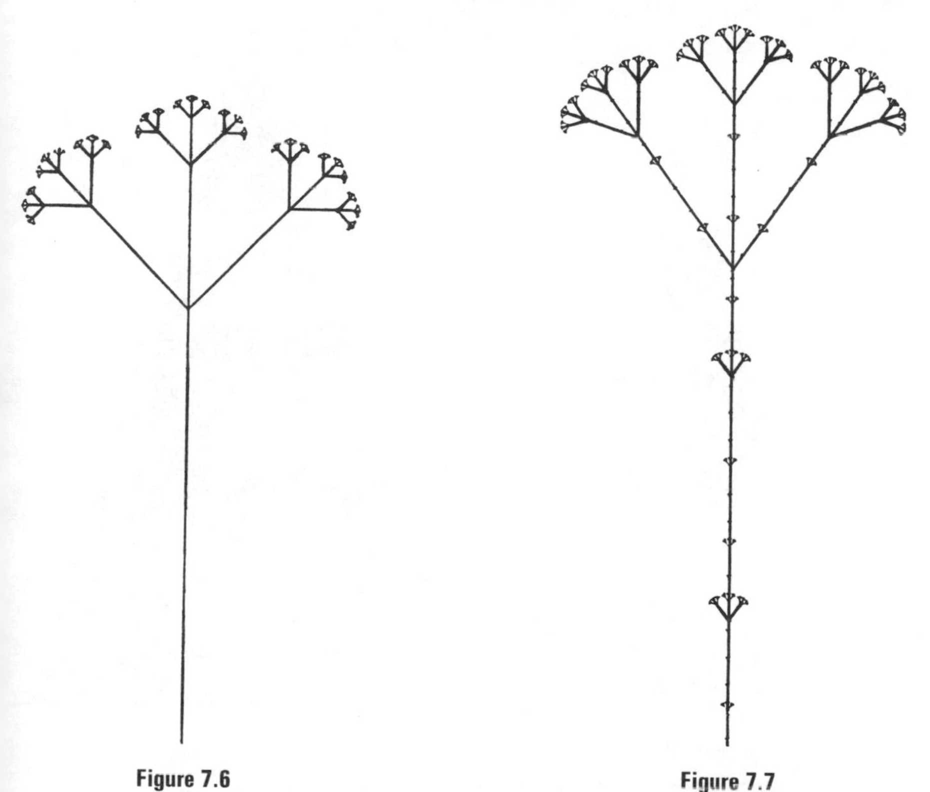

Figure 7.6 Figure 7.7

A third example is of two triangles A and B with the angles $\pi/5$, $2\pi/5$, $2\pi/5$ and $\pi/5$, $\pi/5$, $3\pi/5$ of Figure 7.8. Triangle A breaks up into two triangles similar to A and one triangle similar to B, while triangle B breaks up into a triangle similar to A and one similar to B (reduction by $\frac{1}{2}(\sqrt{5}-1)$). This observation may be exploited to obtain the aperiodic pavings of Penrose (see Figure 7.9 and [4] and [5] for the details).

A second generalization of the notion of self-similarity which fits in naturally

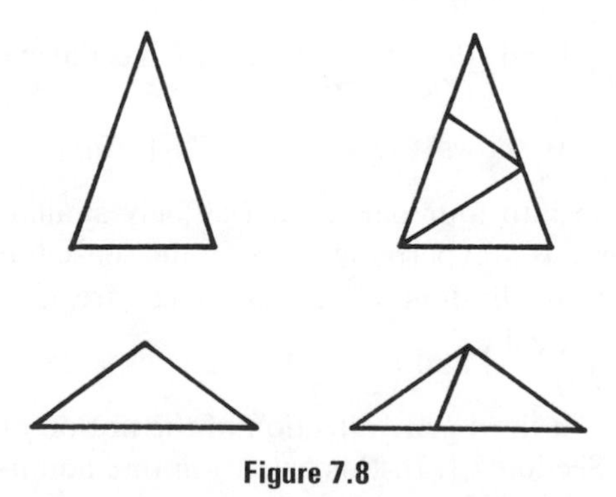

Figure 7.8

Figure 7.9

with the formalism of the last section is that of *self-affinity* (and multi-type self-affinity in due course): one requires that the parts F_j may be obtined from F by means of a linear transformation followed by a translation. A simple example is given by the Kieswetter curve, described by $S=\{a,b\}$, $\theta(a)=abbb$, $\theta(b)=baaa, f(a)=\begin{pmatrix}1\\1\end{pmatrix}$ and $f(b)=\begin{pmatrix}1\\-1\end{pmatrix}$. Here, L has the matrix representation $\begin{pmatrix}4 & 0\\0 & -2\end{pmatrix}$. Figure 7.10 shows the curves $L^{-n}K[\theta^n(a)]$, for $n=1,\ldots,5$. The limit curve breaks up into four parts, each affinely similar to the whole. We observe that this curve is a hyperbolic curve in the sense introduced by Dubuc [7]. The linear parts of the four affine mappings are given by the matrices $\begin{pmatrix}\frac{1}{4} & 0\\0 & -\frac{1}{2}\end{pmatrix}, \begin{pmatrix}\frac{1}{4} & 0\\0 & \frac{1}{2}\end{pmatrix}, \begin{pmatrix}\frac{1}{4} & 0\\0 & \frac{1}{2}\end{pmatrix}$ and $\begin{pmatrix}\frac{1}{4} & 0\\0 & \frac{1}{2}\end{pmatrix}$.

We finally consider a third generalization of the notion of self-similarity and of the formalism of Section 7.1. In this extension one admits several linear (or

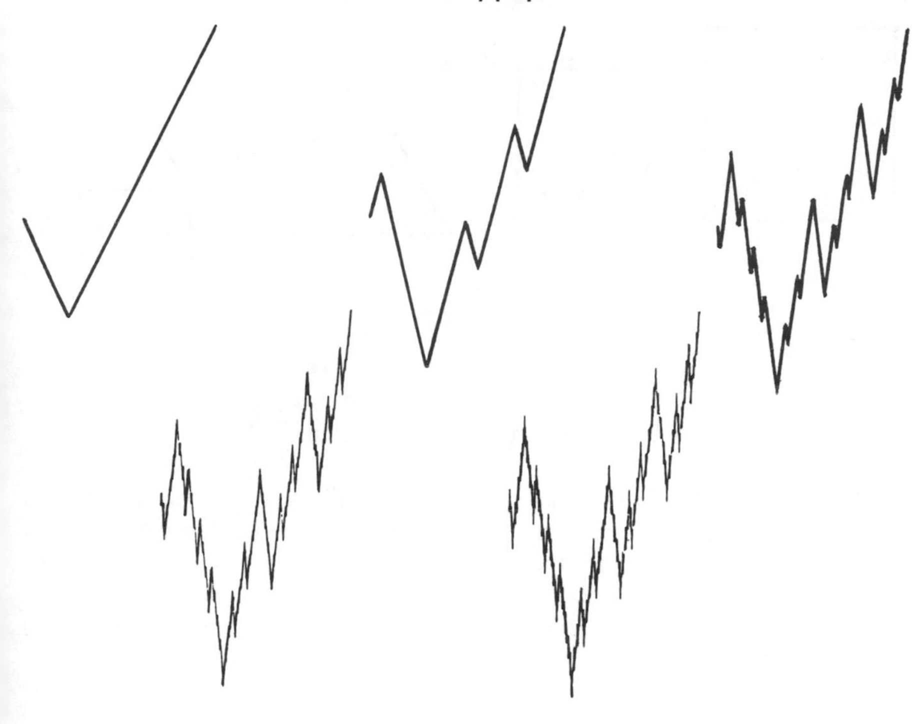

Figure 7.10

even more general) transformations in the same set. Some particular cases have already been considered by Mandelbrot [9], Marion [10], Hutchinson [8] and Dubuc [7]. Let S be an alphabet and $\theta: S^* \to S^*$ a substitution. Suppose that there exists a mapping $f: S^* \to R^d$ and a mapping $L: (\mathbf{R}^d)^S \to \mathbf{R}^d)^S$ such that, for each $s \varepsilon S$,

$$f(s) = \sum_{j=1}^{m} L^{-1}(t_j)(f(t_j)) \qquad \text{if } \theta(s) = t_1 \cdots t_m$$

We follow the treatment of Bedford [1]. Let the mapping $F: (\mathscr{K}(\mathbf{R}^d))^S \to (\mathscr{K}(\mathbf{R}^d))^S$ be given by

$$F(K)(s) = \bigcup_{j=1}^{m} \left\{ L^{-1}(t_j)(K(t_j)) + \sum_{i<j} L^{-1}(t_j)(f(t_i)) \right\}$$

if $\theta(s) = t_1 \cdots t_m$ and $K = (K(s))_{s \in S} \in (\mathscr{K}(\mathbf{R}^d))^S$. One then shows that F is a contraction on $(\mathscr{K}(\mathbf{R}^d))^S$ (for the product of the Hausdorff metrics on each factor) if all the $L^{-1}(s)$ are Lipschitz, with constant less than 1 and fixed point the origin. If at least one of the factors $K(s)$ is non-empty for an essential letter s, $F^n(K) \to K_\theta$ is a unique non-trival fixed point of F. If the $L(s)$ are linear, we shall say that K_θ is *piecewise self-affine*.

As an example, take $S = \{c_\alpha^i, t_\alpha^i, v_\alpha^i : i = 0, 1, 2, 3; \alpha \varepsilon [0, 1]\}$ with $f: S^* \to \mathbf{R}^2$ defined

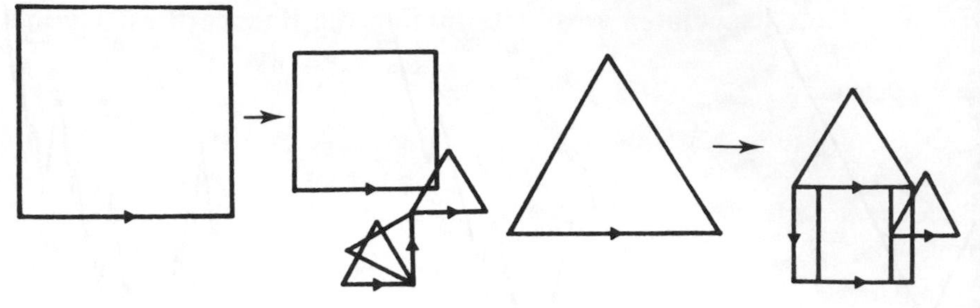

Figure 7.11

by

$$f(c_\alpha^i) = f(t_\alpha^i) = f(v_\alpha^i) = \left(\cos\frac{2\pi i}{4}, \sin\frac{2\pi i}{4}\right)$$

and θ defined by $\theta(c_\alpha^0) = v_{1/9}^0 v_{1/9}^1 c_{2/3}^0 v_{1/9}^2 v_{4/9}^3 t_{1/3}^0 t_{1/3}^1 t_{1/3}^0$ (independent of α), with $\theta(c_\alpha^1), \ldots, \theta(c_\alpha^3)$ being defined symmetrically; $\theta(t_\alpha^0)$ can be worked out from the diagrams in Figure 7.11 and, finally, $\theta(v_\alpha^i) = v_1^i$ for each i and $\alpha\varepsilon$ [0.1]. The mapping L is simply given by $L(S_\alpha^i)(x) = \alpha x (x \varepsilon \mathbf{R}^2)$ for $s = c, t, v; i = 0, 1, 2, 3$ and $\alpha\varepsilon$ [0, 1]. Figure 7.12 shows the component $K(C_1^0)$ of $F^5(K)$, where K consists of squares, triangles and sets which are empty for the v_α^i.

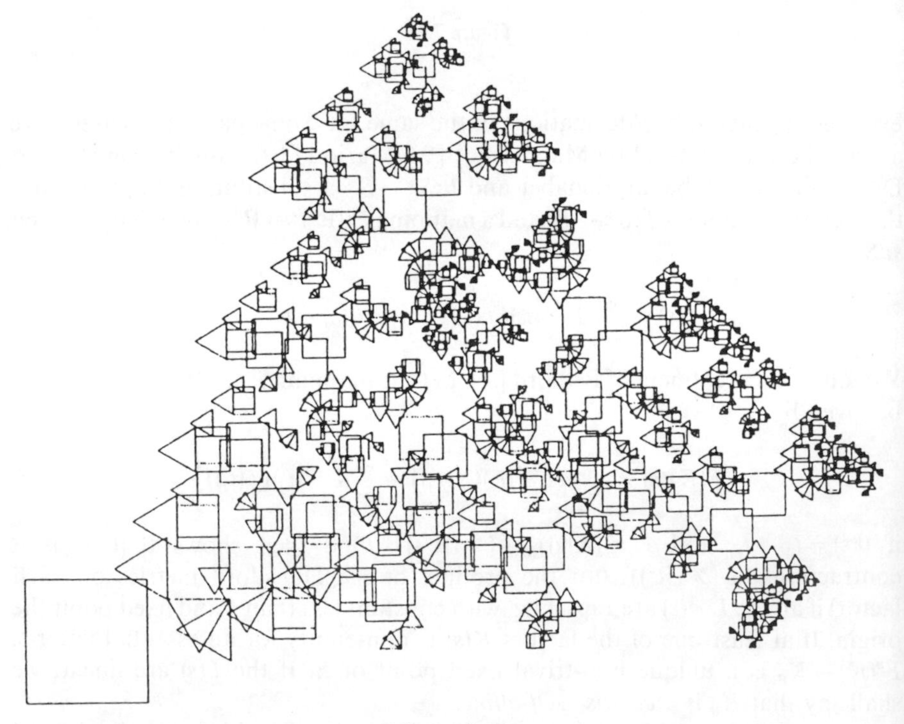

Figure 7.12

7.3 Problems associated with determining the Hausdorff dimension

7.3.1 Solubility

Let us return to the situation of Thorem 1, so that we consider a fractal curve K_θ determined by an alphabet S, a substitution $\theta: S^* \to S^*$ and an expansive representation $L: \mathbf{R}^d \to \mathbf{R}^d$. Constructing economical coverings of K_θ one easily finds that

$$(**) \qquad \dim K_\theta \leqslant d - 1 + \frac{\log \lambda_{\max} - \sum_{j=1}^{d-1} \log |\lambda_j|}{\log |\lambda_d|}$$

where $|\lambda_1| \leqslant \cdots \leqslant |\lambda_d|$ are the absolute values of the proper values of L and $\lambda_{\max}$ is the greatest proper value, whose existence is guaranteed by the Perron–Frobenius theorem, of the matrix $M = (m_{st})_{s,t\varepsilon S}$, where m_{st} is equal to the number of times that t appears in $\theta(s)$.

The problem that arises is under what conditions does one have equality in (**)?

An initial difficulty is actually fairly trivial. If the substitution θ does not possess an irreducibility property, one could generate very different fractals with the same θ. In order to avoid this situation, we will henceforth suppose that the substitution θ is *essentially mixing*, i.e. there exists an integer N such that each essential letter s occurs in each word $\theta^N(t)$, for t essential.

A second obstacle to having equality in (**) is that the Hausdorff dimension of K_θ does not only depend on the proper values of L in the case where L is not a similarity. We return to this in the following section.

In the case where θ is essentially mixing and L is a similarity there remains a final obstacle of a geometric kind: if the curve crosses itself too much, the upper bound (**) will be too great. We call the fractal $K_\theta = K_\theta(W)$ soluble if

$$\liminf_{n \to \infty} \frac{m_d(K\varepsilon[\theta^n(W)])}{|\theta^n(W)|_E} > 0$$

where $m_d(\cdot)$ denotes the Lebesgue measure on $\mathbf{R}^d$, $A^\varepsilon\{x\varepsilon\mathbf{R}^d : \|x - a\| < \varepsilon, a\varepsilon A\}$ for $A \subset \mathbf{R}^d$, and $|W|_E$ is the number of essential letters in a word W. Note that the upper bound (**) simplifies in the case of a similarity:

$$\dim \mathrm{K}_\theta \leqslant \frac{\log \lambda_{\max}}{\log |\lambda_d|}$$

Theorem 3 ([1,4]) *Let* K_θ be generated by an essentially mixing substitution θ and a representation L, where L is a similarity with proper values of modulus $|\lambda|$. Let $\lambda_{\max}$ be the greatest proper value of M, restricted to the essential letters. Then $\dim \mathrm{K}_\theta = \log \lambda_{\max}/\log |\lambda|$ if and only if K_θ is soluble.

Note that solubility is equivalent to the 'open set condition' of Hutchinson [8] and that conditions or algorithms for checking solubility are not known in general (for some special cases, see [3] and]6]). Nevertheless, it is clear that

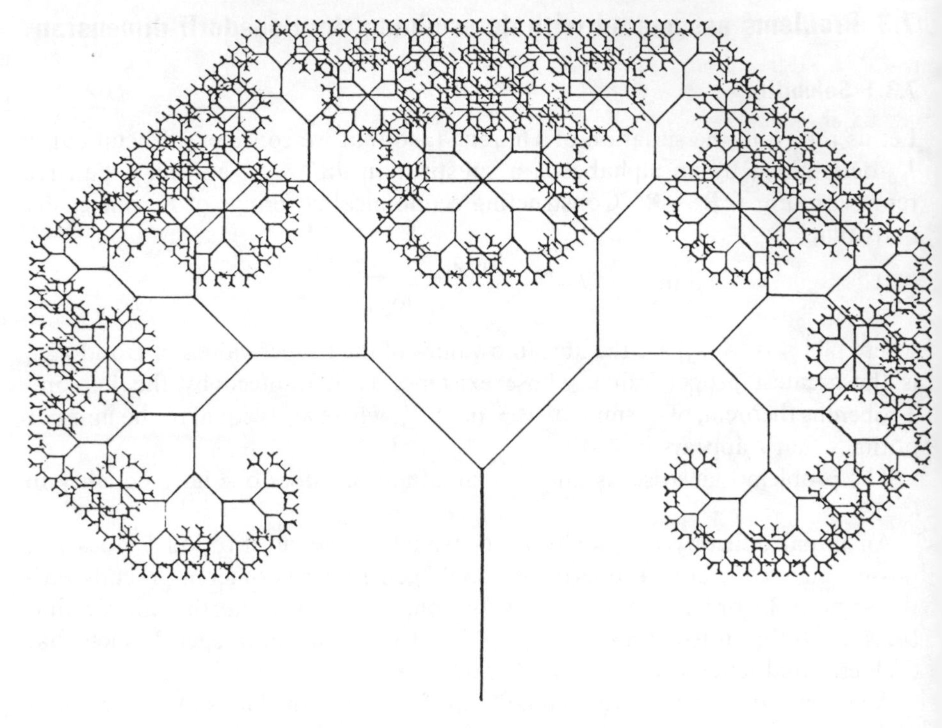

Figure 7.13

for fractals constructed like the Cantor set, one does have solubility. Figure 7.13 shows a fractal (an optimal embedding of the binary tree in $\mathbf{R}^2$, whose dimension is equal to 2) for which there are further obstacles to demonstrating solubility.

7.3.2 Self-affine fractals

We are going to consider soluble fractals, but with L an arbitrary linear mapping. One does not then necessarily have equality in the estimate

$$(**) \qquad \dim K_\theta \leqslant d - 1 + \frac{\log \lambda_{\max} - \sum_{j=1}^{d-1} \log |\lambda_j|}{\log |\lambda_d|}$$

As an example, we consider the cartesian product of two Cantor sets. For $p \geqslant 3$, let C_p be the p-adic Cantor set:

$$C_p = \{x \varepsilon [0,1] : x = \sum x_n p^{-n}, x_n = 0 \text{ or } x_n = p-1\}$$

We assert that one can generate the product $C_p \times C_q (p \geqslant q \geqslant 3)$ by means of a substitution. For this, take $S = \{a, b, c, d, \bar{a}, \bar{b}, \bar{c}, \bar{d}\}$ and let θ be defined on S^* by

$$\theta(a) = ab\bar{a}^{p-2}da, \qquad \theta(b) = bc\bar{b}^{q-2}ab, \qquad \theta(c) = cd\bar{c}^{p-2}bc$$

$$\theta(d) = da\bar{d}^{q-2}cd, \qquad \theta(\bar{a}) = \bar{a}^p, \qquad \theta(\bar{b}) = \bar{b}^p$$

$$\theta(\bar{c}) = \bar{c}^p, \qquad \theta(\bar{d}) = \bar{d}^p$$

(We here write $s^p = ss\cdots s$ (p times) for $s\varepsilon S$.) Let $f: S \to \mathbf{R}^2$ be given by $f(a) = f(\bar{a}) = -f(c) = -f(\bar{c}) = \begin{pmatrix} 1 \\ 0 \end{pmatrix}$ and $f(b) = f(\bar{b}) = -f(d) = -f(\bar{d}) = \begin{pmatrix} 0 \\ 1 \end{pmatrix}$. $K[s]$ is chosen to be polygonal for $s = a, b, c, d$ and the empty set otherwise. Then θ admits an expansive representation L matrix $\begin{pmatrix} p & 0 \\ 0 & \bar{q} \end{pmatrix}$, and one verifies that $C_p \times C_q = K(abcd) = K_{p,q}$. The estimate given us

$$\dim K_{p,q} \leqslant 1 + (\log 4 - \log 9)/\log p$$

while Hausdorff proved that

$$\dim C_p \times C_q = \dim C_p + \dim C_q = \log 2/\log p + \log 2/\log q$$

which is always strictly less than the above estimate unless $p = q$.

There nevertheless exist self-affine fractals for which one has equality in (**), e.g. example 4.7 in [2] or the Kieswetter curve which has dimension $\frac{3}{2}$ [1].

We now present a recent result due to Bedford for a special case in which the complexity of the problem is already apparent. Let p, q be two integers satisfying $p \geqslant q \geqslant 2$. Let K_1 be the subset of $[0,1]^2$ defined by taking the union of a selection of rectangles of the form $[i/p, (i+1)/p] \times [j/q, (j+1)/q]$ $(i\varepsilon\{0,\ldots,p-1\}$, $j\varepsilon\{0,\ldots,q-1\})$, as, for example, in Figure 7.14 (where $p = 5$, $q = 4$).

Let K be the self-affine fractal (the linear mapping being associated with the proper values p and q) obtained by replacing each rectangle of K_1 by an affine copy of K_1, etc. Let K_j be the number of rectangles with a side of the form $[j/q, (j+1)/q]$ (in the example of Figure 7.14, $k_0 = 2$, $k_1 = 0$, $k_2 = 3$ and $k_4 = 1$). Let

$$\mathscr{P} = \left\{ \alpha - (\alpha_0, \ldots, \alpha_{q-1}) \colon \sum_{j=0}^{q-1} \alpha_j = 1 \text{ and } \alpha_j = 0 \text{ if } k_j = 0 \right\}$$

Theorem 4 ([1])

$$\dim K = \max_{\alpha \in \mathscr{P}} \left\{ \frac{-\sum_{j=0}^{k-1} \alpha_j \log \alpha_j}{\log \mathrm{q}} + \frac{\sum_{j=0}^{q-1} \alpha_j \log \mathrm{K}_j}{\log p} \right\}$$

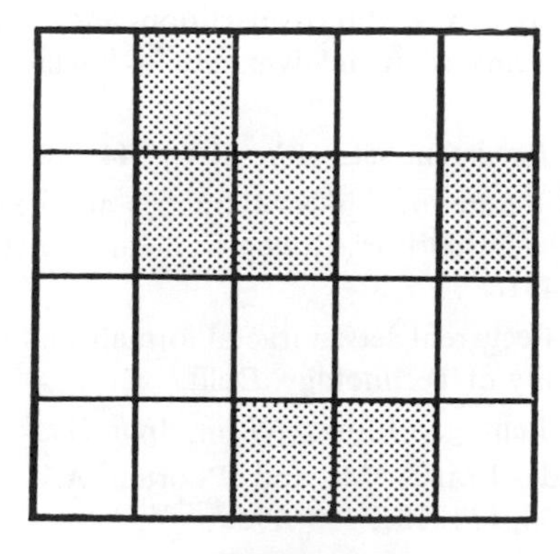

Figure 7.14

Recall that the capacity (or logarithmic density) of a bounded set $A \subseteq \mathbf{R}^d$ is defined by

$$\operatorname{cap}(A) = \lim_{\varepsilon \downarrow 0} \frac{\log N(\varepsilon)}{-\log \varepsilon}$$

(if this limit exists), where $N(\varepsilon)$ is the smallest number of balls of radius ε needed to cover A. It is always the case that $\dim A \leqslant \operatorname{cap} A$. As Bedford [1] remarks, there is a widely held belief that the capacity of a compact set is equal to its dimension. The sets K of Theorem 4 provide simple counter-examples to this. If t is the number of J's for which $K_j \neq 0$, then [1]

$$\operatorname{cap}(K) = \frac{\log t}{\log q} + \frac{\log(1/t \sum_{j=0}^{q-1} k_j)}{\log p}$$

7.3.3 Piecewise self-affine fractals

It is observed in [4] that the dimension of the piecewise self-affine fractal at the end of the Section 7.3 (see Figure 7.12) must be the (unique) positive solution of the equation

$$\left[1 - \left(\frac{2}{3}\right)^{\alpha}\right]\left[1 - \left(\frac{1}{3}\right)^{\alpha} - \left(\frac{5}{9}\right)^{\alpha}\right] - 3\left(\frac{1}{3}\right)^{\alpha} 2\left(\frac{4}{9}\right)^{\alpha} = 0$$

(The fractal is so formed as to be soluble). The question of the dimension of piecewise self-similar fractals is studied by Bedford [1]. His results confirm the above conjecture. He finds a general answer to the question, using ideas of Rufus Bowen on the notion of a variational principal for the pressure of a continuous function on a dynamical system.

References

1. Bedford T. (1984) Crinkly curves, Markov partitions and dimension, Thesis, Warwick University. Also: (1985) Dimension and dynamics for fractal recurrent sets, Cambridge Report, Cambridge.
2. Dekking F. M. (1982) Recurrent sets, *Adv. Math.*, **44**, 78–104.
3. Dekking F. M. (1982) Iterated morphisms, fractals and generalised random walks, in *Fête des Mots. Acte des Journées*, Rouen, 7–8 June, GRECO de programmation du CNRS (1983), pp. 4–16.
4. Dekking F. M. (1982) Recurrent sets: a fractal formalism, Report 8232, Dept. Math. and Inf., Delft University of Technology, Delft.
5. Dekking F. M. (1983) Pentagon tilings, *Nieuw Arch. Wisk.*, **1** (4), 63–69.
6. Dekking F. M. Mendès-France M. and Poorten A.V.D. (1982) Folds! *Math. Intelligencer*, **4**, 130–138, 173–181, 190–195.

7. Dubuc S. (1982) Une foire des courbes sans tangentes, *Actualités Mathématiques. Actes du VIème Congres du GMEL*, Gauthier Villars, paris.
8. Hutchinson J. E. (1981) Fractals and self-similarity, *Indiana Univ. Math. J.*, **30**, 713–747.
9. Mandelbrot B. B. (1982) *The Fractal Geometry of Nature*, W. H. Freeman, San Francisco.
10. Marion J. (1979) Le calcul de la mesure de Hausdorff des sous ensembles parfaits isotypiques de $\mathbf{R}^n$. *C. R. Acad. Sci., Paris*, **289**, 65–68.

Chapter 8 Introduction to packing measures and dimensions

J. Peyrière
Department of Mathematics, University of Paris-sud

8.1 Introduction

Since Hausdorff first defined a fractional dimension, many other notions have been introduced, corresponding to various ways of measuring the 'thickness' of a set, and, depending on the problem at hand, one or other of these notions turns out to be natural in the context. Here, we are going to study the packing dimension recently defined by C. Tricot and we shall compare it with the Hausdorff and Bouligand dimensions.

The crucial role played by the fractal dimension in numerous natural phenomena was discovered by B. Mandelbrot and rather than paraphrasing his book *The Fractal Geometry of Nature*, I urge the reader to study it for himself. In that book, also, the method of determining the dimension depends on the phenomenon being studied, and, in general, this dimension corresponds to a model of the phenomenon that is valid in a range of orders of magnitude. We give two examples.

Consider a square slab of insulating material on which a mass m of metal has been deposited by evaporation (Kapitulnik: Pb; Voss: Au). One finds that the deposit is not regular and one is interested in the conductivity between two opposite sides of the slab. It turns out that, below a critical threshold value m_c of the mass, the conductivity is zero, while above it, the slab is conducting. Study of the conductivity near m_c reveals a critical exponent related to the fractal dimension of the deposit corresponding to $m = m_c$. Here is how this dimension is determined. The slab, assumed to have a side of unit length, is divided into 2^{2n} squares of side 2^{-n}, and N_n denotes the number of these squares on which there is a metallic deposit. A graph is plotted of $\log N_n$ against $n \log 2$. One finds that, for a large range of values of n, the points lie on a straight line whose slope is the looked-for fractal dimension. Of course, for very large values of n, corresponding to squares of side less than the atomic diameter, the behaviour of N_n will be different. Mathematically, the situation is as follows: strictly speaking, the mass in question is a finite set, the number of whose points is of the order of the Avogadro number. It therefore has dimension zero. The dimension found above is the Bouligand dimension, to be defined in due course, of a set corresponding to a continuous (non-discrete) model valid between two scales differing by a large number of orders of magnitude.

For a second example we consider catalysis. Pfeiffer has obtained a fractal dimension in the following way. The number N_ε of molecules, approximately spheres of radius ε which are adsorbed by the catalyst, is measured. For a sufficiently large range of values of ε, one finds that the points $(\log 1/\varepsilon, \log N_\varepsilon)$ lie on a straight line whose slope is the dimension sought. As far as we know, this does not correspond to any notion of dimension to be found in mathematics. In fact, if the catalyst is described by the part of space which it occupies, N_ε is the number of disjoint balls of radius ε which touch E, whose interior lies in the complement of E, and a less easily formalized property, which can be brought 'from the outside' without being blocked by means of a certain number of channels. This corresponds neither to a covering nor to a packing (see the definition later on). This is an interesting situation to study from a mathematical point of view.

8.2 Hausdorff measure and dimension

Consider a bounded subset E of euclidean space $\mathbf{R}^3$. Recall how Lebesgue defines the volume of E. For a given positive number ε, consider a covering of E by balls B_j of radius r_j less than ε and form the quantity $\sum_j \frac{4}{3}\pi r_j^3$. This number is greater than the volume we want to define, particularly if there is too much overlapping of the ball, and so we consider, amonst the coverings, those that are 'economical'. In other words, one considers $m_\varepsilon(E) = \inf\{\sum_j \frac{4}{3}\pi r_j^3\}$, where the infimum is taken over the set of coverings of E by balls of radius less than ε. One then puts $m(E) = \lim_{\varepsilon\to 0} m_\varepsilon(E)$; this is the outer Lebesgue measure which coincides, at least for 'good' sets, with the volume.

Hausdorff measures are obtained by adapting the above construction. Let E be a part of a metric space, euclidean space $\mathbf{R}^d$, for example. Put

$$H_{\alpha,\varepsilon}(E) = \inf\left\{\sum_j (\operatorname{diam} B_j)^\alpha;\ E \subset \bigcup B_j, \operatorname{diam} B_j \leqslant \varepsilon\right\}$$

where α and ε are positive parameters.

Define $H_\alpha(E) = \lim_{\varepsilon\to 0} H_{\alpha,\varepsilon}(E)$, a quantity that can be possibly zero or infinite. When $\alpha = d$, H_α coincides, up to a multiplicative factor, with Lebesgue measure. This quantity H_α is an α-dimensional measure, in the sense that, if the set E' is homothetic to E with a scale factor ξ, we have

$$H_\alpha(E') = \xi^\alpha H_\alpha(E)$$

For a fixed set E, regard $H_\alpha(E)$ as a function of α. One shows that there exists a number α_0 such that if $\alpha < \alpha_0$, then $H_\alpha(E)$ is infinite, while if α is strictly greater than α_0, then $H_\alpha(E)$ is zero. This cut-off value α_0 is the Hausdorff dimension of E, denoted by dim E. One knows nothing, *a priori*, about the number $H_{\alpha_0}(E)$, which can possibly be zero or infinity.

Here is an important property of the Hausdorff dimension: if the set E is a countable union of the family of sets E_n, one has $\dim E = \sup_n \dim E_n$. In particular, since the dimension of a single point is 0, every countable set has Hausdorff dimension zero.

We now show how an upper bound on the Hausdorff dimension may be given. Let $N_\varepsilon(E)$ denote the least number of balls in a covering of E by balls of radius ε. It follows from the definition of $H_{\alpha,\varepsilon}(E)$ that

$$H_{\alpha,\varepsilon}(E) \leqslant (2\varepsilon)^\alpha N_\varepsilon(E)$$

and hence

$$H_\alpha(E) \leqslant \varliminf_{\varepsilon \to 0} 2^\alpha \varepsilon^\alpha N_\varepsilon(E)$$

and

$$\dim E \leqslant \varlimsup_{\varepsilon \to 0} \frac{\log N_\varepsilon(E)}{\log 1/\varepsilon}$$

A further dimensional index $\Delta(E) = \varlimsup_{\varepsilon \to 0}(\log N_\varepsilon(E)/\log 1/\varepsilon)$ is thus revealed. This index is the Bouligand or Minkowski dimension, or the logarithmic density. As we have just seen, we have $\dim E \leqslant \Delta(E)$. The inequality can be strict, as witness the set $Q \cap [0, 1]$, the set of rational numbers lying between 0 and 1, which has Hausdorff dimension zero and logarithmic density equal to 1.

The difference between the Hausdorff dimension and the logarithmic density lies in the fact that in one case one uses coverings by equal balls, while in the other, more complicated coverings are allowed. In both cases, the set E is approached from the outside. In the next section we are going to adopt a sort of dual approach, from the inside.

8.3 Packing measures and dimension

The definitions and results which follow are due to C. Tricot and to C. Tricot and S. J. Taylor. They are to be found in C. Tricot's thesis (Orsay, 1983).

Let E be a subset of a metric space $\mathbf{R}^d$ for example. A packing of E is a collection of pairwise disjoint balls contained in E. Consider, for two positive numbers α and ε, the quantity

$$\Lambda_{\alpha,\varepsilon}(E) = \sup\left\{\sum_j (\operatorname{diam} B_j)^\alpha\right\}$$

where $\{B_j\}$ is a packing of E by balls of radius $< \varepsilon$. Put

$$\Lambda_\alpha(E) = \lim_{\varepsilon \to 0} \Lambda_{\alpha,\varepsilon}(E)$$

One can then show that, if $\Delta(E)$ denotes the logarithmic density, then

(a) $\alpha < \Delta(E)$ implies $\Lambda_\alpha(E) = \infty$.
(b) $\alpha > \Delta(E)$ implies $\Lambda_\alpha(E) = 0$.

In other words, starting with Λ_α, just as one starts with H_α to obtain the Hausdorff dimension, one recovers the logarithmic density. Let us define a further quantity:

$$\Lambda'_\alpha(E) = \inf\left\{ \sum_{n=1}^{\infty} \Lambda_\alpha(E_n) : E \subseteq \bigcup_{n=1}^{\infty} E_n \right\}$$

and consider the cut-off value α_0 associated with Λ'_α, where α_0 is defined by the following two properties:

(a) $\alpha < \alpha_0$ implies $\Lambda'_\alpha(E) = \infty$.
(b) $\alpha > \alpha_0$ implies $\Lambda'_\alpha(E) = 0$.

The number α_0 is called the packing dimension of E and is denoted by $\operatorname{Dim} E$, while $\Lambda_\alpha(E)$ is the α-dimensional packing measure of E. We have $\dim E \leqslant \operatorname{Dim} E \leqslant \Delta(E)$, and all the inequalities can be strict.

One of the essential differences between Δ and Dim is that, if E is a countable union of sets E_n, we have $\operatorname{Dim} E = \sup_n \operatorname{Dim} E_n$, while the statement is false with Δ.

Here is a property that shows why the packing dimension is of some interest. Consider two sets E and F and their cartesian product $E \times F$, i.e. the set of pairs (x, y) such that x belongs to E and y to F. The following string of inequalities, each one of which can be strict, holds:

$$\dim E + \dim F \leqslant \dim E \times F \leqslant \dim E + \operatorname{Dim} F \leqslant \operatorname{Dim} E \times F \leqslant \operatorname{Dim} E + \operatorname{Dim} F$$

The above leads to an interesting definition: a set F is said to be regular if $\dim F = \operatorname{Dim} F$. Then, if F is regular, one has $\dim E \times F = \dim E + \dim F$, for every set E. This condition is the least restructive known at present, which ensures that the dimension of a cartesian product is the sum of the dimensions of its factors.

8.4 A method for determining dim and Dim

We start by defining what one means by measure. It is a function μ which assigns a non-negative number to certain subsets of $\mathbf{R}^d$ in such a way that

$$\mu\left(\bigcup_{n=1}^{\infty} A_n\right) = \sum_{n=1}^{\infty} \mu(A_n)$$

whenever the sets A_n are disjoint. This is the mathematical notion describing a distribution of mass.

We have the following result. Suppose that μ is a measure on $\mathbf{R}^d$ and E is a subset of $\mathbf{R}^d$ such that

(a) $\mu(E) > 0$.
(b) For every x in E, $\log \mu(B_r(x))/\log r$ tends, as r tends to 0, to a limit λ independent of $x(B_r(x)$ denotes the ball centre x and radius r). We then have $\dim E = \operatorname{Dim} E = \lambda$.

Thus, whenever the dimension of E is found in this way, one finds at the same time that this set E is regular in the sense already defined.

8.5 The case of brownian trajectories

We start by generalizing what has gone before. Instead of considering expressions of the form $\sum_j (\text{diam } B_j)^\alpha$, we employ sums of the form $\sum_j h(\text{diam } B_j)$, where h is a continuous increasing function with $h(0)$ equal to zero. One obtains, by the same method as before, two measures for E:

$$H(E) = \liminf_{\varepsilon \to 0} \left\{ \sum_j h(\text{diam } B_j) \colon E \subset \bigcup B_j, \text{ diam } B_j \leqslant \varepsilon \right\}$$

and, similarly, $\Lambda'(E)$, when considering packings.

If $H(E)$ is finite and non-zero, one says that h is the function adapted to E, in the sense of Hausdorff. Similarly, if $\Lambda'(E)$ is finite and non-zero, h is said to be adapted to E, in the sense of packings.

C. Tricot and S. J. Taylor have obtained the following result: if E is a brownian trajectory in $\mathbf{R}^d (d \geqslant 3)$, the function $x^2/\log(\log x)$ is adapted to E, in the sense of packings, while $x^2 \log(\log x)$, is adapted in the sense of Hausdorff. This statement is a little more precise than the assertion that $\dim E = \text{Dim } E = 2$.

Chapter 9 Some remarks on the Hausdorff dimension

J.-L. Jonot
Research Group in Membrane Biophysics, University of Paris VII

On a metric sapce (E, d), the following two notions of dimension can be defined:

(a) The topological dimension, denoted by $\dim_T E$
(b) The Hausdorff dimension, denoted by $\dim_H E$

Intuitively, one can interpret these two dimensions for a physical system (φ) described by a metric space (E, d) in the following way:

(a) $\dim_T E$ is the smallest number of parameters needed to describe the system (φ).
(b) $\dim_H E$ is the smallest non-negative real number for which one can define a volume form on E which is not identically zero, such a volume form being entirely described by the metric d of E. 'The real number $\dim_H E$ is therefore the smallest positive real number for which one can construct non-zero measures on the system (φ). Equally, it is the greatest real number for which one can construct finite measures on the system (φ).'

We recall some definitions and notation concerning the Hausdorff dimension.

An integer $N \geqslant 1$ is chosen and $\mathbf{R}^N$ is endowed with its usual topology. The Borel tribe of $\mathbf{R}^N$ is the tribe generated by the topology of $\mathbf{R}^N$. A Borel subset of $\mathbf{R}^N$ is a subset of $\mathbf{R}^N$ which is in the Borel tribe. $\mathscr{B}_N$ denotes the set of Borel subsets of $\mathbf{R}^N$.

Some notation Let $E \subset \mathbf{R}^N$. The diameter of E denoted by $|E|$ (Figure 9.1) is defined by

$$|E| = \sup_{x, y \in E} \{|x - y|\}$$

where $|x - y|$ is the euclidean norm of the vector $x - y \in \mathbf{R}^N$.

Definition

$$M_\alpha(E, \varepsilon) = \inf\left\{ \sum_{n \in \mathbf{N}} |B_n|^\alpha \colon E \subset \bigcup_{n \in \mathbf{N}} B_n, |B_n| \leqslant \varepsilon \right\}$$

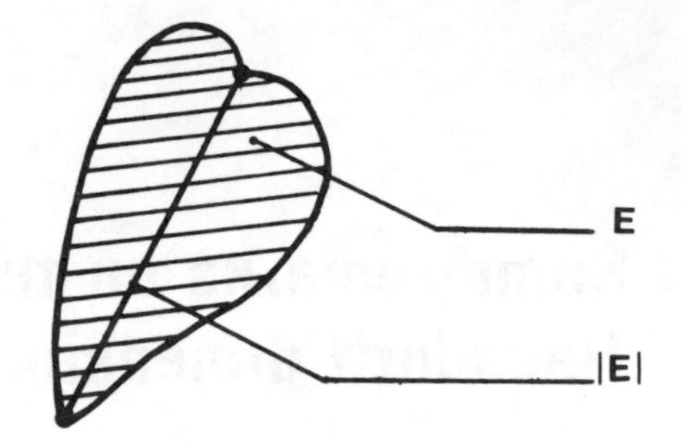

Figure 9.1

Figure 9.2 The B_n cover E and $|B_n| \leqslant \varepsilon$

where $\{B_n\}_{n\in\mathbf{N}}$ is an at most countable family of open balls of $\mathbf{R}^N$ of diameter $|B_N| \leqslant \varepsilon$ (Figure 9.2).

There are several elementary properties associated with this definiton.

Properties

(a) $M_\alpha(E, \varepsilon)$ is an increasing function of ε. Hence $\lim_{\varepsilon\to 0, \varepsilon>0} M_\alpha(E,\varepsilon) = M_\alpha(E)$ exists (possibly $+\infty$).

(b) M_α is an outer measure on the Borel subsets of $\mathbf{R}^N$, i.e. M_α is a mapping of $\mathscr{B}_N$ into $\bar{\mathbf{R}}^+$,

$$M_\alpha\, \mathscr{B}_N \to \bar{\mathbf{R}}^+$$

such that

$$E \to M_\alpha(E)$$

$$M_\alpha\left(\bigcup_{n\in\mathbf{N}} E_n\right) \leqslant \sum_{n\in\mathbf{N}} M_\alpha(E_n)$$

M_α is not in general a measure on $\mathbf{R}^N$, but note that M_α, restricted to a certain Borel subset E of $\mathbf{R}^N$, is a measure on E.

Definition M_α is the α-dimensional Hausdorff measure of $\mathbf{R}^N$.

One can define the Hausdorff dimension of a Borel subset of $\mathbf{R}^N$, and this definition can be extended to an arbitrary metric space (E, d).

Property and definition

(a) If $M_\alpha(E) < \infty$ then $M_\beta(E) = 0$ for $\alpha < \beta$.
(b) We have the following table:

α	0	$\dim E$	∞
$M_\alpha(E)$	∞	$\bar{\mathbf{R}}^+$	0

Definition The Hausdorff dimension of E, denoted by $\dim_H E$, is the non-negative real number defined by

$$\begin{aligned}\dim_H E &= \sup\{\alpha \in \bar{\mathbf{R}}^+ : M_\alpha(E) = \infty\} \\ &= \inf\{\alpha \in \bar{\mathbf{R}}^+ : M_\alpha(E) = 0\}\end{aligned}$$

To define the Hausdorff dimension of E we have covered E by open balls of diameter less than $\varepsilon > 0$. We approach E from the outside. The idea now is to approach the Borel subset E of $\mathbf{R}^N$ from the inside, and we will thus construct a real number, denoted by $\operatorname{Dim}_H E$, which is the 'packing' dimension of E.

Definition Let E be a Borel subset of $\mathbf{R}^N$. Put

$$\mathring{A}_\alpha(E, \varepsilon) = \sup\left\{\sum_{n\in\mathbf{N}} |B_n|^\alpha : \{B_n\}_{n\in\mathbf{N}} \text{ pairwise disjoint, } |B_n| \leqslant \varepsilon, \bar{B}_n \cap \bar{E} \neq \phi, \forall_{n\in\mathbf{N}}\right\}$$

Observe that $\bar{B}_n \cap \bar{E} \neq \phi$ is equivalent to $d(B_n, E) = \inf_{x\in B_n, y\in E} d(x, y) = 0$ (see Figure 9.3). One also says that B_n is at zero distance from E.

We have the following properties, immediate consequences of the definition of $\mathring{A}_\alpha(E, \varepsilon)$.

Figure 9.3 $B_n \cap E = \phi$ but $\bar{B}_n \cap \bar{E} = \{$ 1 point$\}$

Properties

(a) $\mathring{A}_\alpha(E,\varepsilon)$ is a decreasing function of ε, and so $\lim_{\varepsilon\to 0}\mathring{A}_\alpha(E,\varepsilon)$ exists and is denoted by $\mathring{A}_\alpha(E)$ $(0\leqslant \mathring{A}_\alpha(E)\leqslant +\infty)$.

(b) $\mathring{A}_\alpha$ is not outer measure; however, we have the following result:

$$\mathring{A}_\alpha(E_1\cup E_2)\leqslant \mathring{A}_\alpha(E_1)+\mathring{A}_\alpha(E_2)$$

Definition The logarithmic density of a Borel subset of $\mathbf{R}^N$ is defined by

$$\begin{aligned}\Delta(E) &= \inf\{\alpha\in\bar{\mathbf{R}}^+ : \mathring{A}_\alpha(E)=0\}\\ &= \sup\{\alpha\in\bar{\mathbf{R}}^+ : \mathring{A}_\alpha(E)=\infty\}\end{aligned}$$

We have the following property:

Property $\Delta(\bar{E})=\Delta(E)$. In particular Δ is not σ-stable, i.e. we do not have the equality

$$\Delta\left(\bigcup_{n\in\mathbf{N}} E_n\right)=\sup_{n\in\mathbf{N}}\Delta(E_n)$$

for a general countable family of Borel subsets $\{E_n\}_{n\in\mathbf{N}}$ of $\mathbf{R}^N$.

Although $\mathring{A}_\alpha$ is not an outer measure, one can construct an outer measure on E in the following way:

Definition and property Let $q_\alpha(E)=\inf\{\sum_{n\in\mathbf{N}}\mathring{A}_\alpha(E_n): E=\cup_{n\in\mathbf{N}}E_n\}$. Then q_α is an outer measure on $\mathbf{R}^N$ and we have that for every Borel set E in $\mathbf{R}^N$,

$$q_\alpha(E)\leqslant \mathring{A}_\alpha(E)$$

We can define the 'packing' dimension of an arbitrary Borel set in $\mathbf{R}^N$.

Definition The 'packing' dimension of E is

$$\begin{aligned}\mathrm{Dim}_H E &= \inf\{\alpha\in\bar{\mathbf{R}}^+ : q_\alpha(E)=0\}\\ &= \sup\{\alpha\in\bar{\mathbf{R}}^+ : q_\alpha(E)=\infty\}\end{aligned}$$

Property $\dim_H E\leqslant \mathrm{Dim}_H E\leqslant \Delta(E)$.

We give several practical methods for calculating $\Delta(E)$, $\dim_H E$ and $\mathrm{Dim}_H E$.

Property If $M_r(E)$ is the largest number of balls of radius lying between $r/2$ and r which are disjoint and meet $\bar{E}$, then

$$\Delta(E)=\overline{\lim_{r\to 0}}\frac{\mathrm{Ln}(M_r(E))}{\mathrm{Ln}(1/r)}$$

where $\overline{\lim}$ is the limit superior.

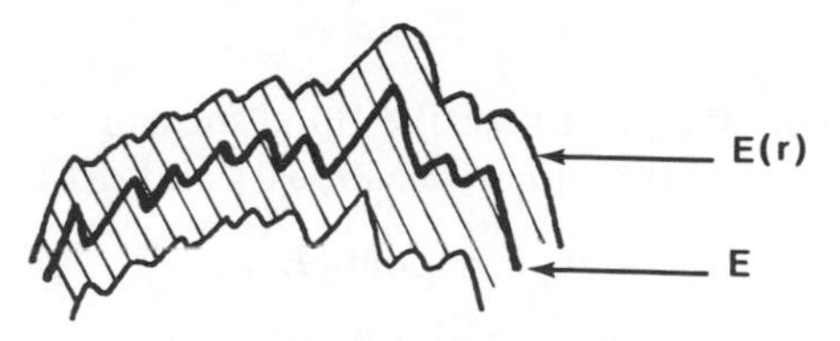

Figure 9.4

Property Let $E(r)=\{x\in\mathbf{R}^N: d(x,E)<r\}$ (see Figure 9.4). Then

$$\Delta(E)=\overline{\lim_{r\to 0}}\left[N-\frac{\mathrm{Ln}(|E(r)|)}{\mathrm{Ln}(r)}\right]$$

where $|E(r)|$ is the diameter of $E(r)$.

As a consequence of these properties we have the following corollary:

Corollary

(a) $\dim E \leqslant \Delta(E)$
(b) $M_\alpha(E,r)\leqslant CM_r(E)r^\alpha$ and $M_\alpha(E)\leqslant \underline{\lim}_{r\to 0}(\mathrm{Ln}(M_r(E))/\mathrm{Ln}(1/r)$ where $\underline{\lim}$ is the limit inferior.

Starting with positive Borel measures on $\mathbf{R}^N$, we are going to show how to calculate $\dim_H E$ and $\mathrm{Dim}_H E$.

Definition Let E be a Borel set and μ a Borel measure on $\mathbf{R}^N$ of finite mass such that $\mu(E)>0$. The set of such measures is denoted by $M_{E_1<\infty}$.

Notation For $\mu\in M_{E_1<\infty}$, we put

$$\varphi\mu(x)=\underline{\lim_{r\to 0}}\frac{\mathrm{Ln}\,\mu(E\cap B_r(x))}{\mathrm{Ln}(r)}$$

$$\phi\mu(x)=\overline{\lim_{r\to 0}}\frac{\mathrm{Ln}\,\mu(E\cap B_r(x))}{\mathrm{Ln}(r)}$$

and

$$\gamma(\mathrm{E})=\sup_{\mu\in M_{E_1<\infty}}\ \inf_{X\in E}\varphi\mu(x)$$

$$\Gamma(E)=\inf_{\mu\in M_{E_1<\infty}}\ \sup_{X\in E}\phi\mu(x)$$

Theorem 1

(a) $\dim_H E=\gamma(E)$
(b) $\mathrm{Dim}_H E=\Gamma(E)$

The previously stated result follows from this.

Properties

(a) If $\dim_H E = \alpha$, then $M_\alpha|_E$ is a positive Borel measure on E.
(b) If $E \subset \{x \in \mathbf{R}^N\colon \operatorname{Ln} \mu(B_r(x))/\operatorname{Ln}(r) \to \alpha \text{ as } r \to 0\}$ and $\mu(E) > 0$, then

$$\dim_H E = \operatorname{Dim}_H E = \alpha$$

In the case where $\dim_H E = \operatorname{Dim}_H E$, the Borel set E is regular, by definition.

In order to define the notion of fractal curve, it is necessary to introduce:

$$P_\alpha(E, \varepsilon) = \sup \left\{ \sum_{n \in \mathbf{N}} |B_n|^\alpha \colon \{B_n\}_{n \in \mathbf{N}} \text{ pairwise disjoint, } |B_n| \leqslant \varepsilon \text{ and } B_n \text{ centred on } E \right\}$$

Definition Put $P_\alpha(E) = \lim_{\varepsilon \to 0, \varepsilon > 0} P_\alpha(E, \varepsilon)$ and $p_\alpha(E) = \inf \{\Sigma_{n \in \mathbf{N}} P_\alpha(E_n)\colon E = \bigcup_{n \in \mathbf{N}} E_n\}$.

Property If $0 < M_\alpha(E) = P_\alpha(E) < +\infty$, then α is a whole number.

Definition If $N = 2$, E is fractal if $0 < M_\alpha(E) = p_\alpha(E)$ and $\dim_H E \neq \operatorname{Dim}_H E$ (E non-regular).

We generalize the above ideas by changing the standard valuation on $\mathbf{R}$. The idea is then to interpret the notion of Hausdorff dimension for the new metric associated with the given valuation. To do this, it is necessary to generalize the notion of the spaces $\mathbf{R}^N$.

9.1 The spaces $\mathbf{R}_v^N$

9.1.1 The valued field $\mathbf{R}_v$

Definition $\mathbf{R}_v$ is the field of real numbers $\mathbf{R}$ endowed with the valuation

$$\mathbf{R} \to \mathbf{R}^+$$
$$x \to |x|^v$$

where v is a real number in the interval $(0, 1]$.

Property For $\mu, v, \varepsilon (0, 1]$, $\mathbf{R}_v^N \approx \mathbf{R}_\mu^N$, where $\approx$ means homeomorphic.

Proof $|x|^\mu \leqslant 1$ implies $|x|^v = (|x|^\mu)^{v/\mu} < 1^{v/\mu} = 1$.

Properties

(a) The distances d_μ and d_v defined by the valuations $|x|^\mu$ and $|x|^v$ are not L-equivalent if $\mu \neq v$.
(b) $\mathbf{R}_\mu \underset{L}{\approx} \mathbf{R}_v$ if and only if $\mu = v$, with $\mu, v \in (0, 1]$ and where $\underset{L}{\approx}$ means L-homeomorphic.

Lemma

(a) If $0<\nu$ there are no strictly positive constants A and B such that

$$A \leqslant |x|^\nu \leqslant B \qquad \forall x \in \mathbf{R}$$

(b) $$\lim_{x\to\pm\infty} \frac{|x|^\nu}{|x|^\mu} = \begin{cases} +\infty & \text{if } \nu>\mu \\ 1 & \text{if } \mu=\nu \\ 0 & \text{if } \nu<\mu \end{cases}$$

(c) $$\lim_{\substack{x\to 0\\ x\neq 0}} \frac{|x|^\nu}{|x|^\mu} = \begin{cases} 0 & \text{if } \nu>\mu \\ 1 & \text{if } \nu=\mu \\ +\infty & \text{if } \nu<\mu \end{cases}$$

9.1.2 Representation in $\mathbf{R}_\nu^2$ of the unit circle

Let $C_\nu = \{(x, y)\in\mathbf{R}^2: d_\nu((x, y), (0,0)) = 1\}$ be the unit circle, i.e.

$$C_\nu = \{(x, y)\in\mathbf{R}^2: |x|^{2\nu}+|y|^{2\nu}=1\}$$

C_ν is symmetrical with respect to the x and y axes (see Figure 9.5).

9.1.3 The normed vector space $\mathbf{R}_\nu^N$

Definition We endow $\mathbf{R}^N$ with one of the following equivalent norms:

$$\|x\|_{2,\nu} = \left(\sum_{i=1}^{N} |x_i|^{2\nu}\right)^{1/2}$$

$$\|x\|_{1,\nu} = \sum_{i=1}^{N} |x_i|^\nu$$

or

$$\|x\|_{\infty,\nu} = \sup_{i=1,2,\ldots,N} |x_i|^\nu, \qquad x=(x_1,\ldots,x_N)\in\mathbf{R}^N$$

We thus have the real normed vector space of dimension $N\in\mathbf{N}^*$ over the valued field $\mathbf{R}_\nu$.

We denote by $\|\ \|_\nu$ one of the equivalent norms on $\mathbf{R}_\nu^N$ and by d_ν the metric associated with this norm:

$$d_\nu(x, y) = \|x-y\|_\nu, \qquad \forall x, y\in\mathbf{R}_\nu^N$$

The diameter of a subset E of $\mathbf{R}_\nu^N$ is written

$$|E|_\nu = \sup_{x,y\in E} d_\nu(x, y)$$

Property $R_\nu^N \approx \mathbf{R}_\mu^N$ ($\approx$ means homeomorphic). The Borel sets of $\mathbf{R}_\nu^N$ and those of $\mathbf{R}_\mu^N$ are the same for $\mu, \nu\in(0, 1]$.

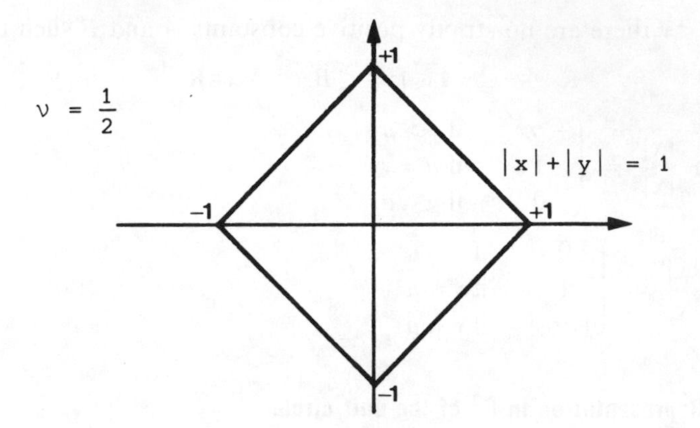

$\nu = \frac{1}{2}$
+1
|x| + |y| = 1
−1
+1
−1

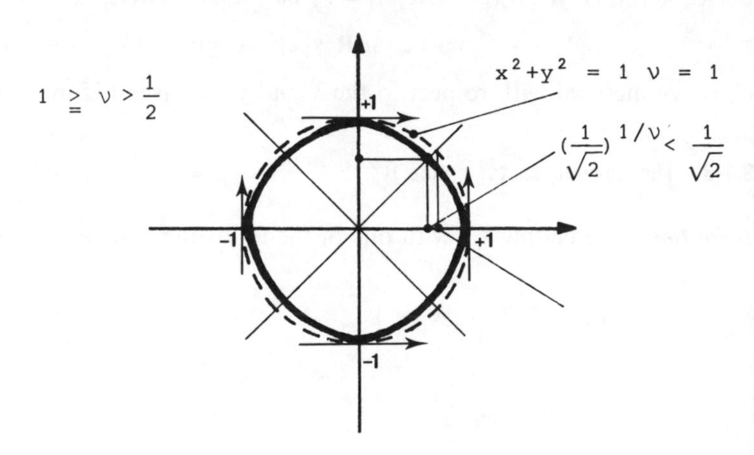

$x^2 + y^2 = 1 \quad \nu = 1$
$1 \geq \nu > \frac{1}{2}$
+1
$(\frac{1}{\sqrt{2}})^{1/\nu} < \frac{1}{\sqrt{2}}$
−1
+1
−1

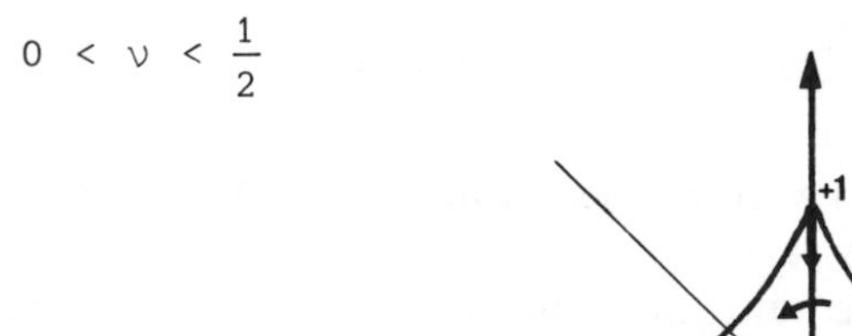

$0 < \nu < \frac{1}{2}$

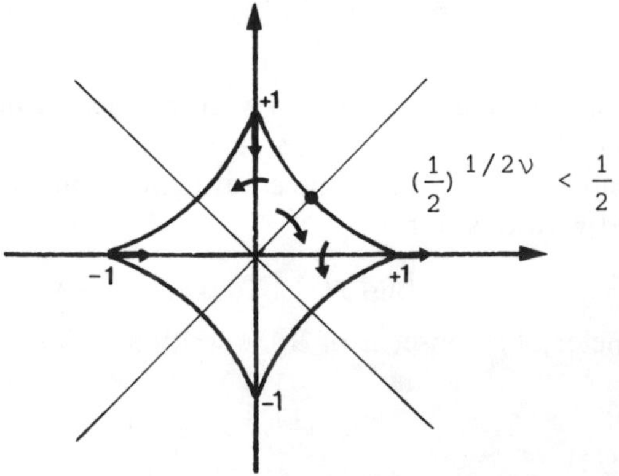

+1
$(\frac{1}{2})^{1/2\nu} < \frac{1}{2}$
−1
+1
−1

Figure 9.5

Property If, our $E \subseteq \mathbf{R}_\nu^N$, one puts $|E|_{\infty,\nu} = \sup_{x,y\in E}\|x-y\|_{\infty,\nu}$, then $|E|_{\infty,\nu} = (|E|_{\infty,1})^\nu$.

Proof for all $x\in\mathbf{R}^N$, $\|x\|_{\infty,\nu} = \sup_{i=1,\ldots,N}|x_i|^\nu = (\sup_{i=1,\ldots,N}|x_i|)^\nu = (\|x\|_{\infty,1})^\nu$ and $t\to t^\nu$ is an increasing mapping $\forall \nu>0$. Hence (cf. curves),

$$\sup_{i=1,\ldots,N} t_i^\nu \leqslant \left(\sup_{i=1,\ldots,N} t_i\right)^\nu, \qquad \forall t_1,\ldots,t_N\geqslant 0$$

$$|x_i| \leqslant \sup_{i=1,\ldots,N}|x_i| = \|x\|_{\infty,1}, \qquad \forall i=1,\ldots,N$$

$$|x_i|^\nu \leqslant (\|x\|_{\infty,1})^\nu, \qquad \forall i=1,\ldots,N$$

$$\sup_{i=1,\ldots,N}|x_i|^\nu = \|x\|_{\infty,\nu} \leqslant (\|x\|_{\infty,1})^\nu, \tag{1}$$

$$\forall i=1,\ldots,N, (|x_i|^\nu)^{1/\nu} = |x_i|$$

$$\sup_{i=1,\ldots,N}|x_i| = \sup_{i=1,\ldots,N}(|x_i|^\nu)^{1/\nu} \leqslant \left(\sup_{i=1,\ldots,N}|x_i|^\nu\right)^{1/\nu}$$

Taking $t_i = |x_i|^\nu$, $\forall i=1,\ldots,N$,

$$(\|x\|_{\infty,1})^\nu \leqslant \|x\|_{\infty,\nu} \tag{2}$$

From (1) and (2), we have: $\|x\|_{\infty,\nu} = (\|x\|_{\infty,1})^\nu$.

Proof of the property

$$|E|_{\infty,\nu} = \sup_{x,y\in E}\|x-y\|_{\infty,\nu} = \sup_{x,y\in E}\{(\|x-y\|_{\infty,1})^\nu\}$$

$$= (|E|_{\infty,1})^\nu$$

If $t_i\geqslant 0$, $\forall i\in I$, then $\sup_{i\in I}\ t_i^\nu = \{\sup_{i\in I} t_i\}^\nu$. Take $I=E\times E$ and $t_{x,y} = \|x-y\|_{\infty,1}$, $\forall (x,y)\in E\times E$.

Lemma $\forall i\in I$, $\sup_{i\in I}\{t_i^\nu\} = \{\sup_{i\in I} t_i\}^\nu$.

Proof of Lemma

$$t\to t^\nu \text{ is increasing from } \mathbf{R}^+\to\mathbf{R}^+, \qquad \forall \nu>0$$

$$t_i \leqslant \sup_{i\in I} t_i, \qquad \forall i\in I$$

$$t_i^\nu \leqslant \left(\sup_{i\in I} t_i\right)^\nu, \qquad \forall i\in I$$

$$\sup_{i\in I}(t_i^\nu) \leqslant \left(\sup_{i\in I} t_i\right)^\nu$$

Conversely

$$t_i = (t_i^\nu)^{1/\nu}$$

$$\sup_{i\in I} t_i = \sup_{i\in I} (t_i^\nu)^{1/\nu} \leqslant \left(\sup_{i\in I} t_i\right)^{1/\nu}$$

$$\left(\sup_{i\in I} t_i\right)^\nu \leqslant \sup_{i\in I} (t_i^\nu)$$

Remark $\mathbf{R}_\nu^N$ is complete ($\mathbf{R}_\nu^N \approx \mathbf{R}_1^N = \mathbf{R}^N$).

9.2 Hausdorff dimension of a subset of $\mathbf{R}_\nu^N$

Definition Let (E, d) be a metric space, and $\mathscr{B}(E)$ the set of bounded subsets of E; i.e. $A \in \mathscr{B}(E)$ if and only if $|A| = \sup_{x,y\in A}\{d(x, y)\} < +\infty$.

A dimension indicator on (E, d) is a mapping α: $\mathscr{B}(E) \to \mathbf{R}^+$ such that:

(a) $E_1 \subset E_2$ implies $\alpha(E_1) \leqslant \alpha(E_2)$.
(b) α is invariant for a class $\mathscr{C}$ of bounded sets.
α is stable if $\alpha(E_1 \cup E_2) = \sup(\{\alpha(E_1), \alpha(E_2)\}$ and α is σ-stable if $\alpha(\cup_{n\in\mathbf{N}} E_n) = \sup_{n\in\mathbf{N}}\{\alpha(E_n)\}$.

Definition Let h be a continuous increasing mapping of $\mathbf{R}^+ \to \mathbf{R}^+$ such that:

(a) $h(\lambda t) \leqslant Ch(t), \lambda \geqslant 1$
(b) $h(0) = 0$.

Standard example $h(t) = t^\alpha, \alpha > 0$.

Now let E be a Borel subset of $\mathbf{R}^N$. Put

$$\mathscr{M}_h^\nu(E, \varepsilon) = \inf\left\{\sum_{n\in\mathbf{N}} h(|B_n|_\nu) \colon |B_n|_\nu \leqslant \varepsilon, E \subset \bigcup_{n\in\mathbf{N}} B_n\right\}$$

The infimum is taken over all coverings $\{B_n\}_{n\in\mathbf{N}}$ of E of diameter less than ε (i.e. $\forall n\in\mathbf{N}, |B_n|_\nu \leqslant \varepsilon$), B_n being a non-empty open subset of $\mathbf{R}^N$, of diameter $|B_n|_\nu \leqslant \varepsilon$.

Property If $|E|_{a,\nu} = \sup_{x,y\in E} \|x - y\|_{a,\nu}$, $a = \infty, 2, 1$ and

$$\mathscr{M}_h^{\nu,a}(E, \varepsilon) = \inf\left\{\sum_{n\in\mathbf{N}} h(|B_n|_{a,\nu}) \colon |B_n|_{a,\nu} \leqslant \varepsilon, E \subseteq \bigcup_{n\in\mathbf{N}} B_n\right\}$$

then we have the following inequalities:

$$\mathscr{M}_h^{\nu,\infty}(E, \varepsilon) \leqslant \mathscr{M}_h^{\nu,2}(E, \varepsilon) \leqslant \mathscr{M}_h^{\nu,1} \leqslant C\mathscr{M}_h^{\nu,\infty}(E, \varepsilon/N)$$

where C is the constant such that

$$h(Nt) \leqslant Ch(t), \qquad \forall t$$

If $h(t)=t^\alpha$, then $C=N^\alpha$.

Proof

$$\|x\|_{\infty,\nu}\leqslant\|x\|_{2,\nu}\leqslant\|x\|_{1,\nu}\leqslant N\|x\|_{\infty,\nu}$$

$$|E|_{\infty,\nu}\leqslant|E|_{2,\nu}\leqslant|E|_{1,\nu}\leqslant N|E|_{\infty,\nu}$$

$$h(|E|_{\infty,\nu})\leqslant h(|E|_{2,\nu})\leqslant h(|E|_{1,\nu})<h(N|E|_{\infty,\nu})\leqslant Ch(|E|_{\infty,\nu})$$

as h is increasing and $N\geqslant 1$. We have

$$E\subseteq\bigcup_{n\in\mathbf{N}}B_n$$

$$\sum_{n\in\mathbf{N}}h(|B_n|_{\infty,\nu})\leqslant\sum_{n\in\mathbf{N}}h(|B_n|_{2,\nu})\leqslant\sum_{n\in\mathbf{N}}h(|B_n|_{1,\nu})$$

$$\leqslant C\sum_{n\in\mathbf{N}}h(|B_n|_{\infty,\nu})$$

whence

$$\mathscr{M}_h^{\nu,\infty}(E,\varepsilon)\leqslant\mathscr{M}_h^{\nu,2}(E,\varepsilon)\leqslant\mathscr{M}_h^{\nu,1}(E,\varepsilon)\leqslant C\mathscr{M}_h^{\nu,\infty}(E,\varepsilon/N)$$

Property $\varepsilon\to\mathscr{M}_h^\nu(E,\varepsilon)$, $\varepsilon>0$, is a decreasing function of ε, and so $\lim_{\varepsilon\to 0,\varepsilon>0}\mathscr{M}_h^\nu(E,\varepsilon)$ exists for any Borel subset of $\mathbf{R}_\nu^N$, and we write

$$\mathscr{M}_h^\nu(E)=\lim_{\substack{\varepsilon\to 0\\ \varepsilon>0}}\mathscr{M}_h^\nu(E,\varepsilon)=\sup_{\varepsilon>0}\mathscr{M}_h^\nu(E,\varepsilon)\in\bar{\mathbf{R}}^+$$

Property If E is a Borel subset of $\mathbf{R}^N$, then

$$\mathscr{M}_h^{\nu,\infty}(E)\leqslant\mathscr{M}_h^{\nu,2}(E)\leqslant\mathscr{M}_h^{\nu,1}(E)\leqslant C\mathscr{M}_h^{\nu,\infty}(E)$$

where C is the constant associated with h for $\lambda=N$.

Remarks

(a) The bounded Borel subsets of $\mathbf{R}_\nu^N$ are the Borel subsets of $\mathbf{R}^N$ because $|E|_{\infty,\nu}=(|E|_{\infty,1})^\nu$ and

$$|E|_{\infty,\nu}\leqslant|E|_{2,\nu}\leqslant|E|_{1,\nu}\leqslant N|E|_{\infty,\nu}$$

(b) If E is an arbitrary subset of $\mathbf{R}_\nu^N$ one can define $\mathscr{M}_h^\nu(E)$ in an analogous way.

Property

$$\mathscr{M}_h^\nu:\mathscr{P}(\mathbf{R}^N)\to\bar{\mathbf{R}}^+$$

satisfies

$$\mathscr{M}_h^\nu\left(\bigcup_{n\in\mathbf{N}}E_n\right)\leqslant\sum_{n\in\mathbf{N}}\mathscr{M}_h^\nu(E_n)$$

and is therefore an outer measure on $\mathbf{R}^N$.

Fundamental property If $h(t)=t^{\alpha}$, $\alpha>0$, we put

$$\mathscr{M}_{\alpha}^{\nu,\infty}(E)=\mathscr{M}_{h}^{\nu,\infty}(E) \quad (*)$$
$$\mathscr{M}_{\alpha}^{\nu,\infty}(E)=\mathscr{M}_{\nu\alpha}^{1,\infty}(E), \qquad \text{for every subset } E\subseteq \mathbf{R}^N$$

Lemma $\mathscr{M}_{\alpha}^{\nu,\infty}(E,\varepsilon)=\mathscr{M}_{\alpha\nu}^{1,\infty}(E,\varepsilon^{1/\nu})$ for every subset E of $\mathbf{R}^N$.

Proof $|B_n|_{\infty,\nu}=(|B_n|_{\infty,1})^{\nu}, \quad \forall n\in N.$

$$\sum_{n\in\mathbf{N}}(|B_n|_{\infty,\nu})^{\alpha}=\sum_{n\in\mathbf{N}}(|B_n|_{\infty,1})^{\nu\alpha}$$

Moreover $|B_n|_{\infty,\nu}\leqslant\varepsilon$, which is equivalent to $|B_n|_{\infty,1}\leqslant\varepsilon^{1/\nu}$ and the result follows.

Remark $\{B_n\}_{n\in\mathbf{N}}$ runs through the set of coverings of E.
Now

$$\lim_{\substack{\varepsilon\to 0\\ \varepsilon>0}}\mathscr{M}_{\alpha}^{\nu,\infty}(E,\varepsilon)=\lim_{\substack{\varepsilon\to 0\\ \varepsilon>0}}\mathscr{M}_{\alpha\nu}^{1,\infty}(E,\varepsilon^{1/\nu})$$

i.e. $(*)$ holds.

Property and definition

(a) If $\alpha<\beta$ and $\mathscr{M}_{\alpha}^{\nu}(E)<\infty$ then $\mathscr{M}_{\beta}^{\nu}(E)=0$.
(b) From (a) the following table can be deduced:

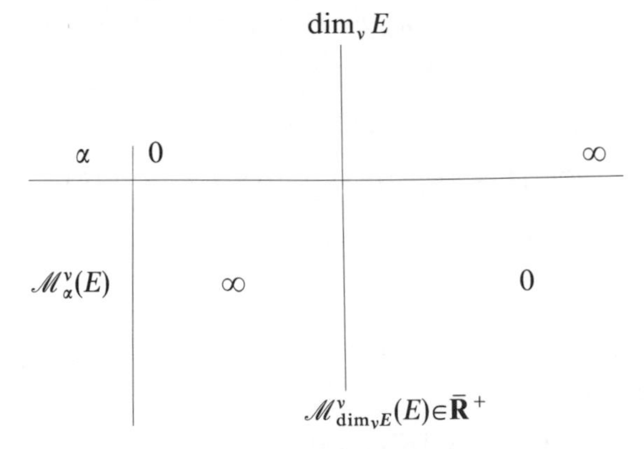

Definition The Hausdorff ν-dimension of $E\subset\mathbf{R}^N$, denoted by $\dim_\nu E$, is the real number defined by

$$\dim_\nu E=\sup\{\alpha\in\bar{\mathbf{R}}^+ : \mathscr{M}_{\alpha}^{\nu}(E)=\infty\}$$
$$=\inf\{\alpha\in\bar{\mathbf{R}}^+ : \mathscr{M}_{\alpha}^{\nu}(E)=0\}$$

Remarks

(a) If $\alpha=\dim_\nu E$ then $\mathscr{M}_{\alpha}^{\nu}(E)\in\bar{\mathbf{R}}^+$.
(b) $\dim_\nu E=\alpha$ does not depend on the equivalent norms chosen, but one does

have

$$\mathscr{M}_\alpha^{\nu,\infty}(E) \leqslant \mathscr{M}_\alpha^{\nu,2}(E) \leqslant \mathscr{M}_\alpha^{\nu,1}(E) \leqslant N^\alpha \mathscr{M}_\alpha^{\nu,\infty}(E)$$

and $\mathscr{M}_\alpha^\nu(E)$ does depend on the norm chosen on $\mathbf{R}_\nu^N$.

Property For any subset E of $\mathbf{R}^N$,

$$\dim_\nu E = \frac{1}{\nu}\dim_1 E$$

Proof

$$\mathscr{M}_\alpha^{\nu,\infty}(E) = \mathscr{M}_{\nu\alpha}^{1,\infty}(E)$$

$$\dim_\nu E = \sup\{\alpha\in\bar{\mathbf{R}}^+ : \mathscr{M}_\alpha^{\nu,\infty}(E) = \infty\}$$

$$= \sup\{\alpha\in\bar{\mathbf{R}}^+ : \mathscr{M}_{\nu\alpha}^{1,\infty}(E) = \infty\}$$

If one puts $\beta = \nu\alpha$, then $\dim_\nu E = (1/\nu)\sup\{\beta\in\bar{\mathbf{R}}^+ : \mathscr{M}_\beta^{1,\infty}(E) = \infty\}$ and $\dim_\nu E = 1/\nu \dim_1 E$.

For a metric space (E, d), we put $\dim E$ for the Hausdorff dimension of E endowed with the metric d. With this notation, we have the following result:

Property

$$\dim_\nu \mathbf{R}^N = \frac{N}{\nu}$$

For $E = \mathbf{R}^N$, $\dim_1 E$ is the Hausdorff dimension of $\mathbf{R}^N$ endowed with the standard metric, and so $\dim_1 E = N$; $\dim \mathbf{R}_\nu^N = \dim_\nu \mathbf{R}^N$ and the required result follows from the preceding property.

Theorem 2 If E is a subset of $\mathbf{R}^N$, there exists a unique $\nu\in(0,1]$ such that $\dim_\nu E = N$. We suppose that the topological dimension of E is not zero.

Proof $\dim_1 E \leqslant N$ $(E\subseteq\mathbf{R}^N)$ and $\dim_1 E \geqslant \dim_T E > 0$, where $\dim_T E$ is the topological dimension of E. There exists a unique $\nu\in(0,1]$ such that

$$0 < \dim_1 E = \nu N \leqslant N, \text{ namely } \nu = \frac{\dim_1 E}{N}$$

Then

$$\dim_\nu E = \frac{1}{\nu}\dim_1 E = \frac{\nu N}{\nu} = N$$

9.3 Volume interpretation of the Hausdorff dimension

Definition If $\mathbf{R}^N$, the open ball, centre x^0 and radius ε, is the set of points of $\mathbf{R}^N$ such that

$$\| x^0 - x \|_\nu^a < \varepsilon; \qquad a = \infty, 2, 1$$

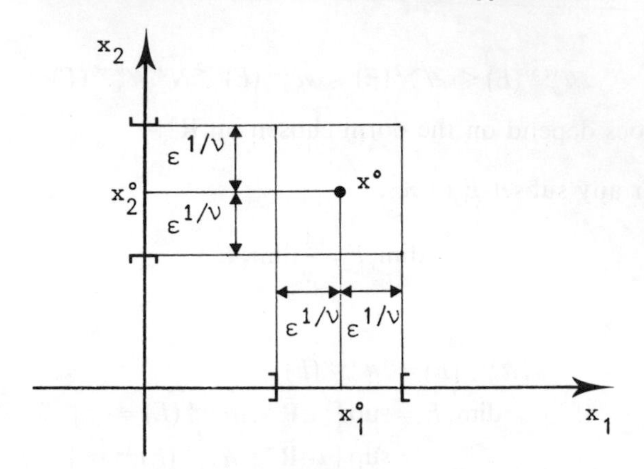

Figure 9.6

If $a=\infty$, $x^0=(x_1^0,\ldots,x_N^0)$ and $x=(x_1,\ldots,x_N)\in\mathbf{R}^N$, then we have, $\forall i=1,\ldots,N$, $|x_i^0-x_i|<\varepsilon^{1/\nu}$, which define a cube centred at x^0 and of side $2\varepsilon^{1/\nu}$.
If $N=2$, Figure 9.6 results.

If $a=2$,

$$\sum_{i=1}^{N}|x_i^0-x_i|^{2\nu}<\varepsilon^2$$

If $a=1$,

$$\sum_{i=1}^{N}|x_i^0-x_i|^{\nu}<\varepsilon$$

(see Figure 9.7), because

$$\sum_{i=1}^{N}|x_i^0-x_i|^{2\nu/2}<(\varepsilon^{1/2})^2$$

Property Let

$$\mathcal{N}_h^{\nu,\infty}(E,\varepsilon)=\inf\left\{\sum_{n\in\mathbf{N}}h(|B_n|_\nu^\infty):|B_n|_\nu^\infty\leqslant\varepsilon,\right.$$

$$\left.E\subseteq\bigcup_{n\in\mathbf{N}}B_n;\{B_n\}\text{ is a family of open balls}\right\}$$

Then

$$\mathcal{N}_h^{\nu,\infty}(E,\varepsilon)=\mathcal{M}_h^{\nu,\infty}(E,\varepsilon)$$

Proof If B_n is an open set of diameter $|B_n|_{\infty,\nu}\leqslant\varepsilon$, there exists an open ball C_n such that (see Figure 9.8)

(a) $B_n\leqslant C_n$

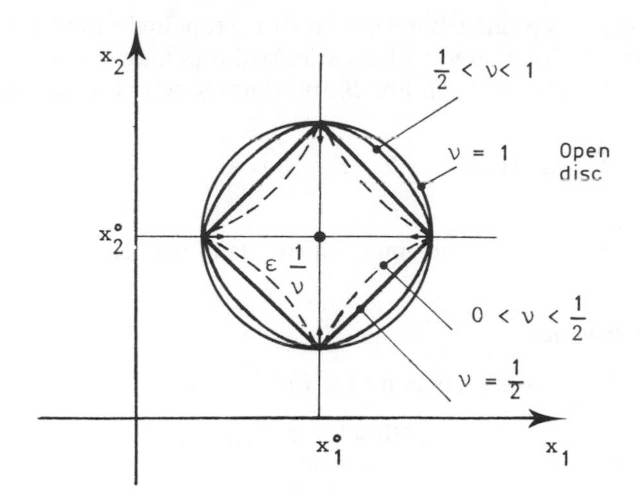

Figure 9.7

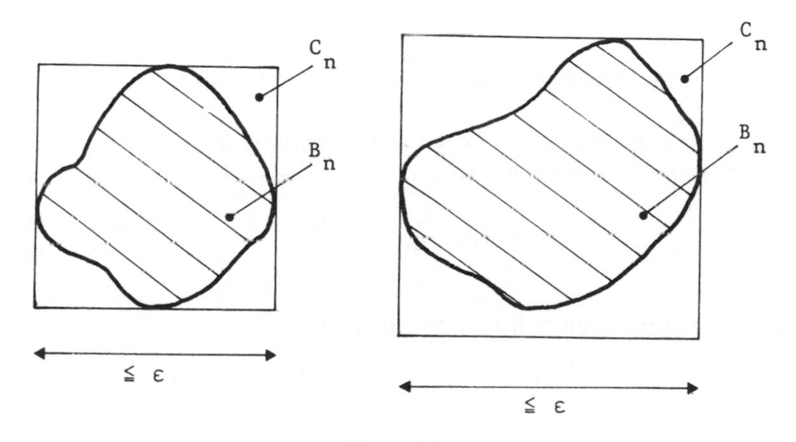

Figure 9.8

(b) $|B_n|_{\infty,\nu} = |C_n|_{\infty,\nu}$

$\{C_n\}_{n\in\mathbf{N}}$ is a covering of E, of diameter

$$|C_n|_{\infty,\nu} = |B_n|_{\infty,\nu} \text{ less than } \varepsilon$$
$$\mathscr{M}_h^{\nu,\infty}(E,\varepsilon) \geqslant \mathscr{N}_h^{\nu,\infty}(E,\varepsilon)$$

The inequality $\mathscr{M}_h^{\nu,\infty}(E,\varepsilon) \leqslant \mathscr{N}_h^{\nu,\infty}(E,\varepsilon)$ is trivial.

Remark In defining the Hausdorff dimension, one need therefore only consider coverings of E by open balls B_n of diameter less than ε.

Notation Let μ be a positive Borel measure on a topological space E (in fact E is a metric space (E, d) endowed with its standard topology) and ν a real number belonging to the interval $(0, 1]$. For B an arbitrary subset of E, put

$$\mu_\nu(B, \varepsilon) = \inf\left\{ \sum_{n\in\mathbf{N}} [\mu(B_n)]^\nu : B \subseteq \bigcup_{n\in\mathbf{N}} B_n; \{B_n\}_{n\in\mathbf{N}} \text{ is a covering of } B \text{ by non-empty open sets of diameter} \leqslant \varepsilon \right\}$$

Property and definition

(a) $\mu_\nu(B, \varepsilon)$ is a decreasing function of ε. Put

$$\mu_\nu(B) = \lim_{\substack{\varepsilon\to 0\\ \varepsilon>0}} \mu_\nu(B, \varepsilon)$$

(b) $\mu_\nu(B)$ is the μ-volume of B of order ν. If $\nu < \omega$ and $\mu_\nu(B) < \infty$ then $\mu_\omega(B) = 0$. Put

$$\begin{aligned}\operatorname{ind}_\mu B &= \sup\{\nu\in(0, 1] : \mu_\nu(B) = \infty\}\\ &= \inf\{\nu\in(0, 1] : \mu_\nu(B) = 0\}\end{aligned}$$

Then $\operatorname{ind}_\mu B$ is the dimension indicator of B for the measure μ.

Theorem 3 Let $\mathbf{R}^N$ be endowed with its standard metric and its Lebesgue measure λ. For each subset $B \subset \mathbf{R}^N$,

$$\operatorname{ind}_\lambda B = \frac{\dim B}{N}$$

Proof For every open ball C_n one can write

$$|C_n|^\alpha = (|C_n|^N)^{\alpha/N} \qquad \text{and} \qquad \sum_{n\in\mathbf{N}} |C_n|^\alpha = \sum_{n\in\mathbf{N}} [\lambda(C_n)]^{\alpha/N}$$

and with the above notation the result follows ($|C_n| = |C_n|_{\infty,1}$).

Remarks

(a) If $\dim B = \alpha$ (the Hausdorff dimension of B) then $\mathscr{M}_\alpha(B) = \lambda_\nu(B)$ where λ is Lebesgue measure on $\mathbf{R}^N$ and ν is the dimension indicator B for Lebesgue measure.
(b) λ_ν is an outer measure of $\mathbf{R}^N$ for $\nu\in(0, 1]$.
(c) $\lambda_\nu(B) \geqslant [\lambda(B)]^\nu$ if B is a Borel subset of $\mathbf{R}^N$.

One can define a dimension indicator on a (non-metric) topological space E endowed only with a positive Borel measure in the following way. For $\nu\in\mathbf{R}^+$ and B an arbitrary subset of E, put

$$\mu^{\nu}(B,\varepsilon)=\inf\left\{\sum_{n\in\mathbf{N}}[\mu(B_n)]^{\nu}:B\subseteq\bigcup_{n\in\mathbf{N}}B_n;\{B_n\}_{n\in\mathbf{N}}\right.$$
$$\left.\text{is a covering of } B \text{ by non-empty open sets such that } \mu(B_n)\leqslant\varepsilon,\ \forall n\in\mathbf{N}\right\}$$

In this definition $\mu^{\nu}(B,\varepsilon)$ is independent of the choice of a metric on E defining the topology of E.

Definition and property

(a) $\mu^{\nu}(B,\varepsilon)$ is a decreasing function of ε. Put

$$\mu^{\nu}(B)=\lim_{\substack{\varepsilon\to 0\\ \varepsilon>0}}\mu^{\nu}(B,\varepsilon)$$

(b) $\mu^{\nu}(B)$ is the μ-volume of B of order ν. If $\nu<\omega$ and $\mu^{\nu}(B)<\infty$ then $\mu^{\omega}(B)=0$. Put

$$\begin{aligned}\operatorname{ind}\mu(B)&=\sup\{\nu\in\bar{\mathbf{R}}^{+}:\mu^{\nu}(B)=\infty\}\\&=\inf\{\nu\in\bar{\mathbf{R}}^{+}:\mu^{\nu}(B)=0\}\end{aligned}$$

Remarks

(a) On a topological space with a measure (E,μ) one can define two invariants independent of the metric defining the topology of E (if E is a metric space), namely:
   (i) The topological dimension of E, which only depends on the topology of E;
   (ii) The dimension indicator of E, which only depends on the structure of the measure space of E, i.e. (E,μ).
(b) If $0\leqslant\nu\leqslant 1$, then $\mu^{\nu}(B)\geqslant[\mu(B)]^{\nu}$ for every Borel set B of E.
(c) If $\nu\leqslant\omega$ then $\mu^{\nu}(B)\leqslant[\mu^{\omega}(B)]\nu/\omega$.

In conclusion, we draw attention to two of the principal features of this work:

(a) The more complicated one makes the metric of the ambient space $\mathbf{R}^N$ of an object E, the more one simplifies the nature of the object E.
(b) On a general (not necessarily metrizable) topological space E, for each Borel measure λ on E, one can define a Hausdorff dimension associated with this Borel measure λ.

Note that changing the valuation on $\mathbf{R}$ means changing the scaling on the standard distance.

Chapter 10 Fractals, materials and energy

Alain Le Méhauté
Compagnie générale d'électricité (Marcoussis)

10.1 Fractals: where are they to be found?

The simplest definition of fractal interfaces will help furnish an answer to the above question: 'They have to do with objects that are as fragmented at the microscopic level as in the large' (Figure 10.1). Thus, objects that fall into no simple category, excepting that under which conceptual problems are just reclassified, are the so-called heterogeneous materials.

Materials made up of several distinct ingredients, for example, can exhibit a fractal structure, e.g. mortars and alloys. Also materials consisting of a single substance in a variety of stable or metastable states:

Solid/solid (polymers or crystalline materials)
Solid/liquid (metals in a fusion bath or saturated solution)
Liquid/gaseous (clouds, aerosols, etc.)

All are characterized by the existence of at least two quite distinct domains. The fractal object is basically an intricate intermingling. We will call 'fractal' the very complex frontier that separates each of two domains. Although it is always difficult to describe the mixture precisely, it is in the existence of this frontier that the thermodynamic significance of the fractal object lies. The set of phase equilibria between the different constituents of the material really rests on this. It is easy to believe that these equilibria can involve the metric of the space directly without any intermediate agent, at least in idcal cases. For this, it is convenient for the indentations to reach right down to the atomic scale. This is indeed the case with defective and syncrystalline crystals.

Until now, there have only been a few studies devoted to these domains and in fact we can only mention two that take account of the concept of fractality: on the one hand, there are studies that have to do with compounds of ionic insertion and, then again, studies dealing with certain highly defective compounds of ionic insertion. Results of experiments in this area have not, for the most part, been published. We can, however, give two relevant examples:

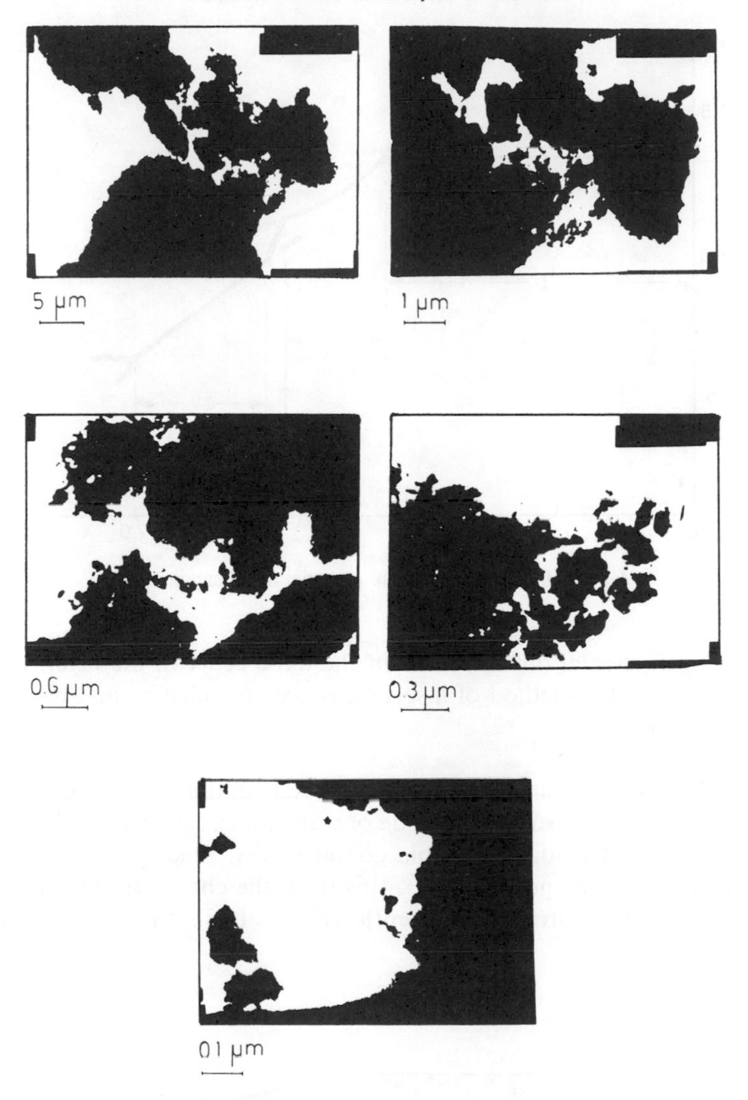

Figure 10.1 Manganese oxide at various enlargements

Example 1. *Dimensional phase transition in Li_xFeS_2* A new type of foliated iron pyrites was synthesized in 1979. It involves a compact hexagonal stacking of sulphur; the iron just occupies the available tetrahedral sites of every other layer. Thus, in the initial state $Li_\varepsilon FeS_2$ ($\varepsilon \to 0$), every other layer has the room to accommodate a lithium ion at a crystal site of the empty layer. As the thermodynamic graph giving the chemical potential of the ion as a function of the concentration (Figure 10.2) shows, this insertion in the layer takes place in two different ways according to whether the amount of ion is greater or less than one. Suitable models (in particular the TEISI model) make it possible to

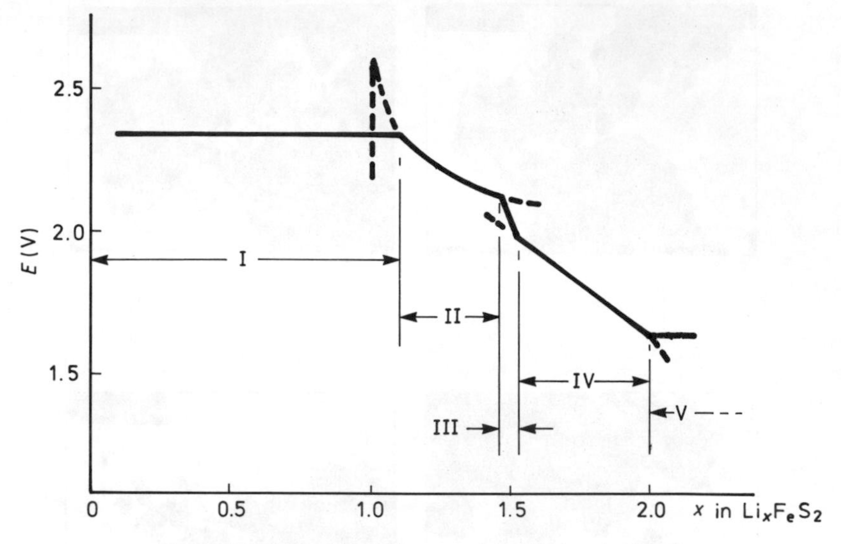

Figure 10.2 Thermodynamic potential of lithium in Li_xFeS_2

go back to the fractal dimension from the ionic order in the layer. One finds that a change in the method of insertion is accompanied by an abrupt change in the fractal dimension.

Example 2. *Another example* Consider a system that does not display the above anisotropic properties: an oxide of manganese, e.g., that used in batteries, which is the most readily available commercially. One undertakes the same experiment as above and finds that, this time, the change in dimension is not brutal but, on the contrary, very 'gentle' (Figure 10.3). This is not the place to

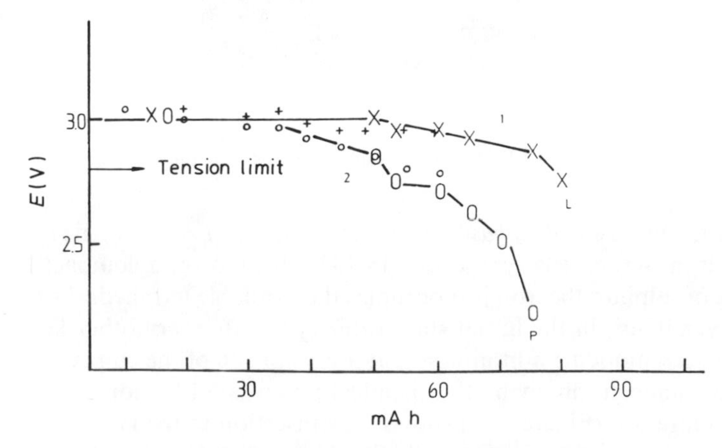

Figure 10.3 Chemical potential of two varieties of manganese oxide (MnO_2) as a function of the fractal dimension

develop the new thermodynamic concepts associated with these observations, but merely to emphasize that these studies have brought to light two facts:

(a) There exists a close relationship between the variance of an open thermodynamical system and its fractal dimension.
    The system can only be univariant in the traditional sense of the term under the condition that its metric be integral.
(b) Biphase systems are not necessarily zero-variant because the metric of the frontier between phases constitutes a degree of freedom by means of which the system may adjust its free energy.

In the examples mentioned, the phase transition exclusively affects the fractal dimension. In the first case, the abrupt transition just shows how different sites of the system are filled without changing anything but the second-order derivatives. In the second case there is a latent heat of change of state and the increase in the fractal dimension is associated with both a diminishing of the free energy of the system and a weakening of its structure. From a crystallographic point of view, the Bragg spots which were originally Dirac functions give place to more complex structures which contain all the information about local disorder and which resemble spots in optics.

Despite this disorder, the concept of phase is valid as long as the frontier between the elements of the mixture can be distinguished. It is enough that it is possible to separate the inside from the outside of bounded domains without regard to the mathmatical analyticity of the frontier.

This limitation leads to the introduction of a geometric entropy which the energy balance-sheet takes account of; the experimental proof of this is provided here. It must again be the case that this frontier has a finite number of double points and be discernable everywhere. The case of indiscernable domains is surely encountered in all cases where the fractal dimension is whole and equal to the topological dimension of the space. In this case, the 'intricate intermingling' property is such that each point of the space must then be regarded as the linking of two phases that are thereafter indiscernable. The environment is then said to be 'homogeneous'. This feature does not necessarily rule out singular types of behaviour with which workers in thermodynamics are familiar. Such behaviour is, in general, formally hidden under the banner of parameters called coefficients of activity. Would it be possible to give a geometric meaning to these parameters? This question will be left open here.

We finally pose two particularly interesting questions:

(a) Is the notion of indiscernability reducible to the notion of complexity?
(b) Can discernable domains persist for an infinitely long time?

The answer to the first question is clearly negative. The notion of complexity is intrinsic and suggests the fundamental non-resolution of the interface due to an infinity of double points which so entangles the curve that the beginning of a 'thread of Ariane' may be impossible to find. In this context, the only outcome of an initial complexity measure of non-uniformity would be a move towards uniformity (the initial system of mixing at the micro level would thus tend to uniformity ... by increase in complexity).

An answer to the second question requires, as a preliminary, that a method of analysis be defined.

10.2 On scales of measurement

A moment's thought on the nature of materials will convince one that:

(a) Unlike mathematical 'objects', physical objects are never 'fractals' in the sense of being a limit as the scale of measurement tends to zero, but are almost surely, in the sense of an overall complexity as a function of scale.
(b) The physical fractal does not necessarily call for a fractal dimension. The 'fractal dimension' can change as a function of the scale of measurement without the physical ideas introduced by Mandelbrot being affected. It seems that this recognition of the generality of the mathematician's ideas is still not widely accepted.
(c) Fractality can be discussed at all bounded levels and without any reference to the Hausdorff dimension.

Despite the manifest differences between the mathematical and physical object, the term 'fractal' is still quite adequate in our opinion. What matters physically is the rule by which one passes from a scale of order n to one of order $n \pm 1$ and the iterative procedure. The limit as n tends to infinity will almost surely be non-fractal and physically one can expect high and low cut-off points outside of which the concept is not appropriate.

It therefore becomes necessary to inquire how a state of thermodynamical equilibrium can be found experimentally to correspond with the fractal dimension taken here in the sense of Mandelbrot. One method would consist of considering the equilibria affecting the local radii of curvature and their distribution. Some very limited efforts in this direction have not yielded anything as yet.

Another method consists of considering a space–time link-up via an exchange process which is either reversible or irreversible and always dissipative. Since such a process never has inifinite velocity, the product of the characteristic exchange velocity with a response or excitation time provides a length that is a candidate for the role of space-gauge.

In the case of thermodynamical equilibrium, the conversion to heat of the carriers of energy (electron, ion, ...) leads to the looked-for characteristic length (a): for example

$$a = \frac{hT}{vm^*}$$

Such an analysis enables one to understand how, thermodynamically, the potential can be sensitive to the metric. This conclusion will be much easier to understand if one recalls that thermodynamic equilibrium can always be regarded as the existence of two thermal interfacial exchange velocities of the same absolute value and in opposite senses.

However, the gauge in most general use is in fact connected with diffusion. In this case, the characteristic gauge-length is given by l with

$$l = \sqrt{Dt}$$

where D, the coefficient of diffusion, is expressed in square centimetres per second. The following three remarks are of particular interest:

(a) According to the diffusion hypothesis, the fractal would only make sense in a quadratic setting (cm^2/s). One could quite reasonably ask why this has to be so?
(b) The necessity of using a process having a finite exchange velocity implies the existence of a functional relationship between reversibility and fractality. The choice of diffusion as the dissipative element is excellent in this respect even if it could be questionable from other points of view. In the thermodynamic case, only the conversion to heat camouflages the dissipation.
(c) Consideration of the dimensional aspects of the diffusion coefficient suggests that one could take account of the role of the gauge in the Fourier space instead of space-time. The reason could lie in the fact that the time transform of a variable is always a time integral of this variable. A space-guage would thus appear naturally upon integration of diffusion-type coefficients.

Remarks (a) and (b) suggest a more general investigation than merely restricting the space-guage to the length of diffusion. Why not, in fact, consider a characteristic length arising from a formula of the type

$$l = (\mathscr{D}t)^{1/d}, \qquad \mathscr{D}[L^d t^{-1}]$$

Going further, why not take as characteristic length

$$\mathscr{F}\left(\frac{\mathrm{d}l}{\mathrm{d}t}\right) = \eta(\omega), \qquad \eta[L]$$

where $\mathscr{F}$ stands for the Fourier transform. A study of the consequences of this choice has led to the so-called TEISI (Transfert d'Energie sur Interface à Similitude Interne) model.

The dissipative processes determining the gauge can then be formulated according to two distinct choices:

(a) Fractal boundary conditions (?) and dissipation take place in the neighbourhood of the fractal (the traditional way) (Figure 10.4).
(b) There are no boundary conditions in the strict sense and the dissipation is determined by the interfacial exchange in a 'fuzzy' interface which confers the metric of the space on the sources of entropy production (TEISI model) (Figure 10.5).

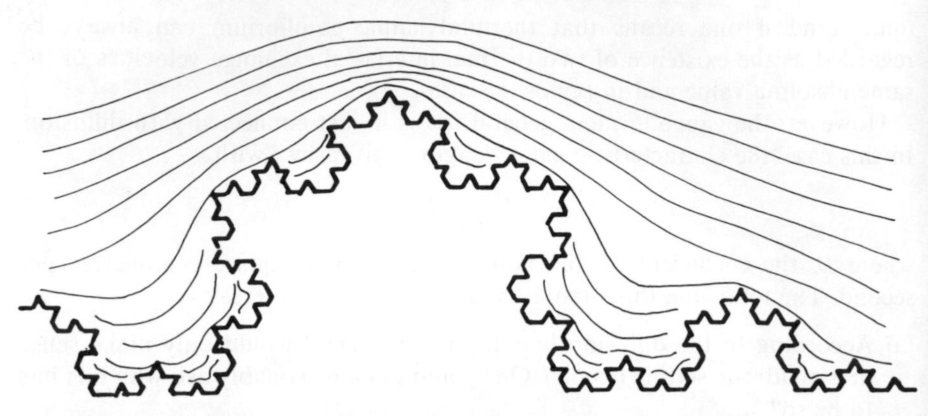

Figure 10.4 Experimental study of the distribution of electric potentials in the neighbourhood of a von Koch interface. It will be noticed that this distribution does not essentially change when the resolution of the interface is increased. The dissipation of the energy occurs in an environment that does not necessarily have the same metric as that of the interface. The relevant 'dimension' is the exchange index, also called the outer dimension

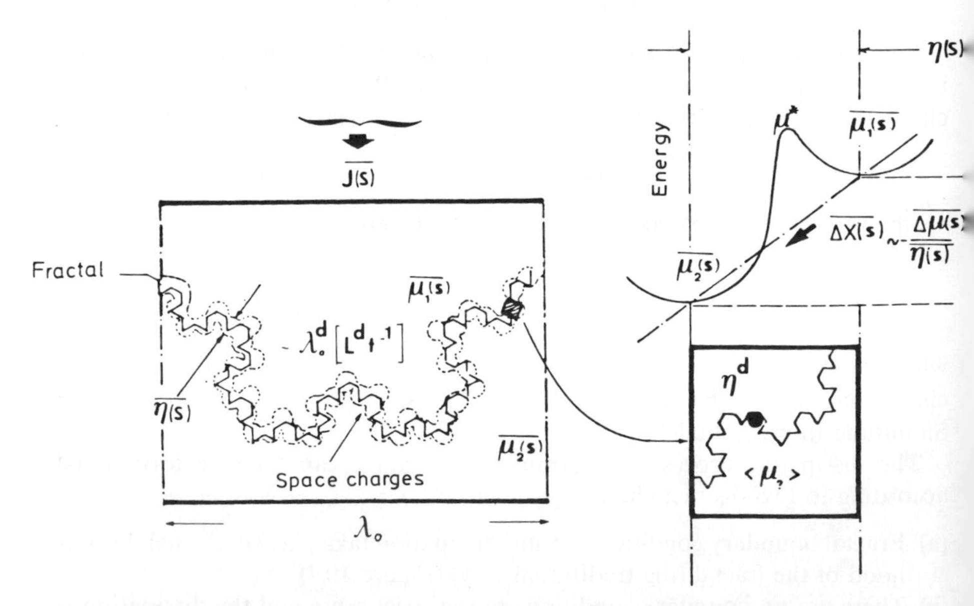

Figure 10.5 Distribution of energy losses according to the TEISI model. Each of the domains on either side of the frontier can be regarded as equipotential except in a thin zone of finite thickness $\eta(S)$ if S is the Laplace variable. The uncertainty is the result of the movement of active elements clinging to the interface. The dissipation will determine a Bouligand-type measure on the interface (cf. TEISI model)

In both cases the dissipative factors are indicative of the metric. However, in the second case this indicator will be direct, so that the exchange kinetics can be a function of the fractal dimension d in the sense of Mandelbrot, while in the first case, the kinetics will always be a function of a function of the dimension d (spectral dimension for example) (cf. Orbach's work).

10.3 The TEISI model

This model adopts the second point of view. It consists of examining objects that are as 'fragmented' on a small scale as in the large, these qualities being in other respects relative (Figure 10.6). One then tries to think of chemical,

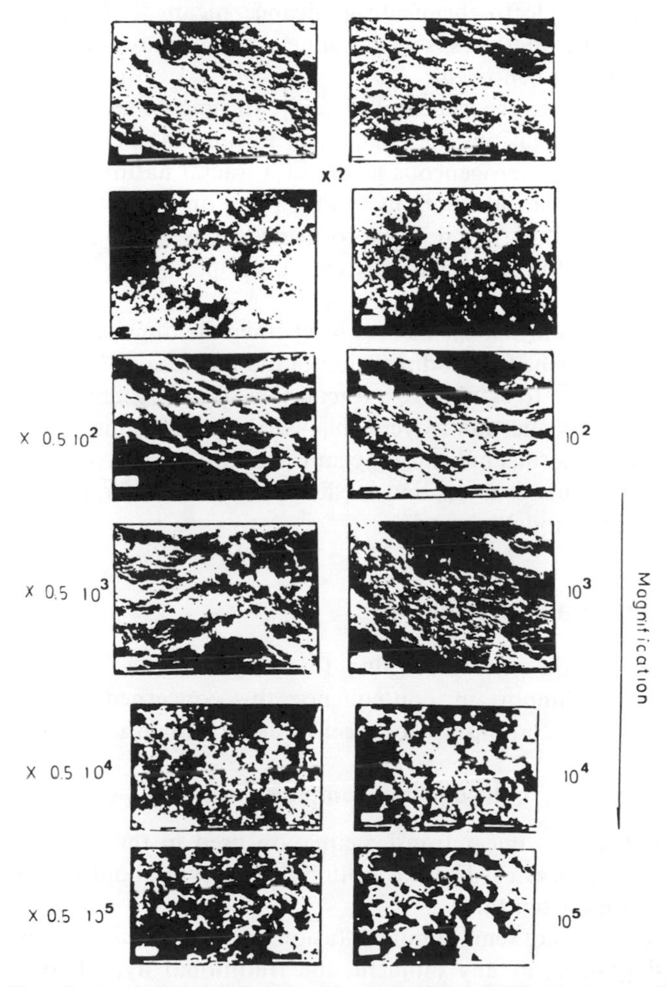

Figure 10.6 The collector of a liquid cathode at various magnifications. What is the magnification of the upper photos?

particularly electrochemical, reactions involving these objects which are such that the sources of entropy production have the same metric as the exchange object. This can easily be achieved by experimental methods using, for example, electrochemical techniques.

The theory then leads one to write the chemical transfer functions in a very singular form. In fact it is shown that the flux–force relationship of the thermodynamics of irreversible processes becomes

$$\mathscr{D}^{1/d-1} \times \text{macroscopic flux} \sim \text{exchange force}$$

Note that $\mathscr{D}^{1/d-1}$ is here a fractional differentiation operator which, in two-dimensional topology, involves the fractal dimension directly.

The dissipative factors considered are exclusively chemical and non-diffusive. This hypothesis, of electrochemical significance, means that the reorganization of molecules at the interfaces is at the same time both faster and less reversible than the transport phenomena.

That this theory is experimentally valid can now be assumed as far as its electrochemical conclusions are concerned. Recall that the idea of the experiments is the use of heterogeneous media of a fractal nature for electrodes and to allow various chemical reactions to take place on these media. The material of the electrode then plays a decisive role since, in as much as there is a unique fractal dimension, the fractional differentiation operator is independent of the kinetics.

The main question that then arises is whether the dimension is equal, as the theory suggests, to the fractal dimension of the medium in Mandelbrot's sense. Having been unable to synthesize a fractal to a specific geometrical pattern, we cannot answer this question as yet. This work thus remains to be done. Note, however, that numerous indicators seem to confirm this hypothesis in the case where the electrochemical reaction is effectively a δ-transfer, as defined according to the TEISI model.

10.4 Fractality and dissipation

The discovery of the generalized flux–force equation had an immediate, and in our view very important, consequence: this concerned the existence of a necessarily close relationship between fractal dimension and irreversible dissipation.

In fact (Figure 10.7), the equation simplifies in two-dimensional topology:

(a) For $d = 1$, to the linear transfer equation used in the thermodynamics of irreversible processes. There is nothing surprising about this as the TEISI model sprang from this.
(b) To the equation of semi-infinite diffusion for $d = 2$. This result is much more surprising since, at any moment, the traditional hypotheses of diffusion models have not been considered (such as, for example, the conservation equations).

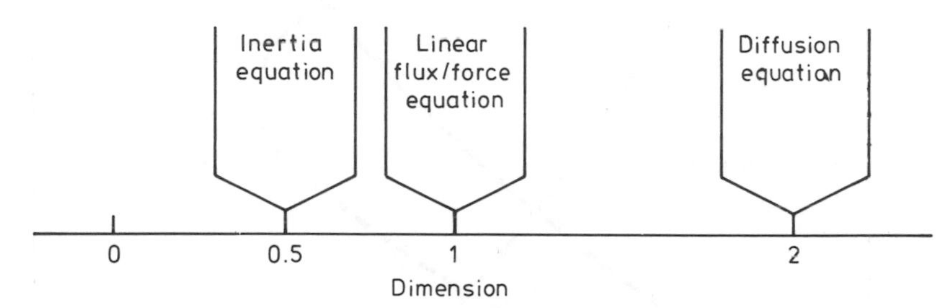

Figure 10.7 Diagram showing the different areas of application of the generalized flux–force equation in a fractal setting

This result thus suggests that the model should be linked with statistical aspects of diffusion (brownian motion, fractional here, and capacity considerations of energy transfer).

(c) For $d < 1$, in particular $d = 0.5$, the dissipative character of these equations tends to diminish. Inertial properties then appear, associated with the emergence of cooperative effects of metric conservation.

This very basic connection between fractal dimension and dissipation is especially interesting in the subject of molecular engineering and process optimization.

For example, it can in fact be shown, according to the most advanced versions of the model, that the energy yield of an irreversible process depends intimately on the fractal dimension of the medium in which it takes place. Thus, the capacity of an electrochemical battery is determined in ideal cases by an entropy invariant which may be written as

$$[\text{Discharge current}]^{d-1} \times [\text{capacity}] = \text{constant}$$

for every value of the capacity less than the theoretical capacity. This law, known ever since batteries have existed and formalized by Peukert in 1891 in connection with the lead/lead oxide accumulator, has its origin in the fractal nature of the electrode which restricts the generator in rapid discharge.

Furthermore, the development of the TEISI model enables one to define methods of renormalization between the discharge curves of batteries suited to different regimes. The specific renormalization techniques depend on the fractal nature of the electrodes (see Figure 10.8).

Optimization of energy storage requires that the dimension of the space be three and not two. Between two and three, the energy function is increasing. On the other hand, optimization of the energy exchanges (power) demands that the dimension of the exchange space be two. It is therefore expedient to settle,

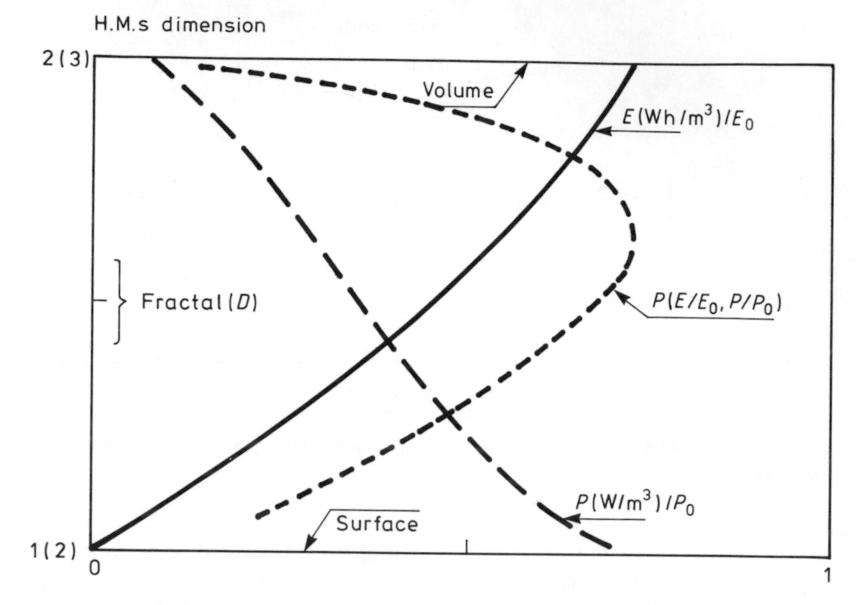

Figure 10.8 Diagrammatic analysis of the dielectric energy (W h)/power (W)

according to the speicific requirements, for an optimum that necessarily calls for a fractal space of dimension lying between two and three. In the case of the lead accumulator, this optimum is in the region of 2.45.

Experiment fully confirms the theoretical results put forward and it should be appreciated that from now on fractal geometry may serve as a tool for the engineer, viz:

(a) An analytical tool for the characterization of fractal interfaces.
(b) An optimization tool whose purpose is to reduce dissipative factors.
(c) Lastly, a tool for synthesis. It is possible, for example, to imagine the synthesis of fractal molecules.

Another topic which almost certainly has connections with fractal characteristics is catalysis. Avnir and Pfeifer, for example, have reexamined the techniques for the characterization of surfaces by the methods of adsorption of molecules of varying sizes. A significant number of materials have now been studied.

10.5 Why should the world be fractal?

As predicted by Mandelbrot, fractality is a property of great generality.

Nevertheless, we still do not know the origin of this state of affairs. There is no hint of a reason to be found in the roots of either the Crystallers laws (geography), the Zipf–Mandelbrot laws (econometrics) or the dimension 2.45 of the lead accumulator.

Figure 10.9 A lead electrode at various magnifications and states of discharge

Furthermore, unlikely resemblances delight in misleading us (Figure 10.9) and suggest the existence of fundamental laws unconcerned with either the geometric scales to which they apply or the dissipative processes of which they are the source or the product.

We are thus left with the following question: Why? We take the risk of attempting a qualitative explanation.

Consider an arbitrary dissipative process beginning at time $t = 0$. This process is a source of entropy which is thereby endowed with the minimal velocity $\sigma(t)$. This entropy is stored in the medium either in the form of 'heat', the traditional form, or in the form of 'information' (disorder) relative to the metric conferred on the medium. In the initial state the metric can be a whole number. In the stationary state the heat produced is entirely dissipated via the calorific capacity of the medium. To accomplish this, the fractal dimension is irreversibly tied to a non-integral value compatible with the irreversible constraints. The bifurcation from one state to another depends on a metric dimensional transition, as can already be seen in Figure 10.8.

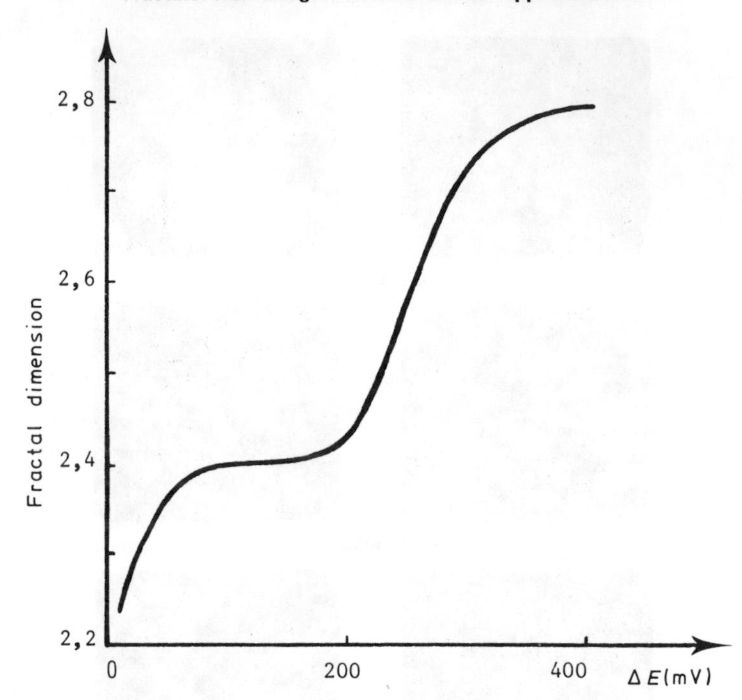

Figure 10.10 Dimensional transition for an electrode sited on a liquid cathode collector

Some people will be able to see, in this rough and ready explanation, the more contemporary picture that one has of turbulence, of micromixtures or of electrical growth. The foundation of these transitions could be sought in the dialectic, very familiar to the engineer, between energy and power (Figure 10.8).

There is no doubt that research in this area has only just begun, but one can bet that dimensional transition far from equilibrium will stimulate some advanced thinking in a generalized theory of phase transitions that has yet to be developed.

Surely this would be a good time to consider phase transitions not only in space-time but in the reciprocal space? Correlations at large distances there become correlations at short distances and the energy factors are expressed there in terms of integrals of actions—all functions likely to reveal the metric.

It is henceforth recognized that the concept of fractality is closely associated with the processes that govern phase transitions, and there thus seems no point in dwelling on them. Nevertheless, we give a pictorial illustration of what could emerge from an analysis in the Fourier space.

Figures 10.11 and 10.12 show the two-dimensional structure of a crystal of quaternane, an imaginary molecule invented by English scientists. The state of this material is given at two distinct temperatures in order to show the phase transitions. It is shown how the domain boundaries, well behaved at low

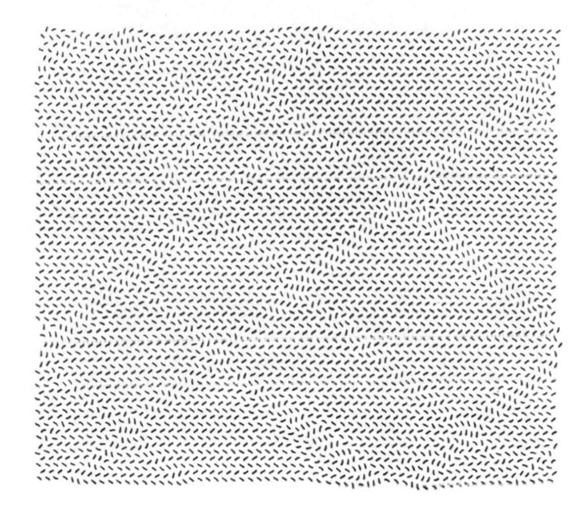

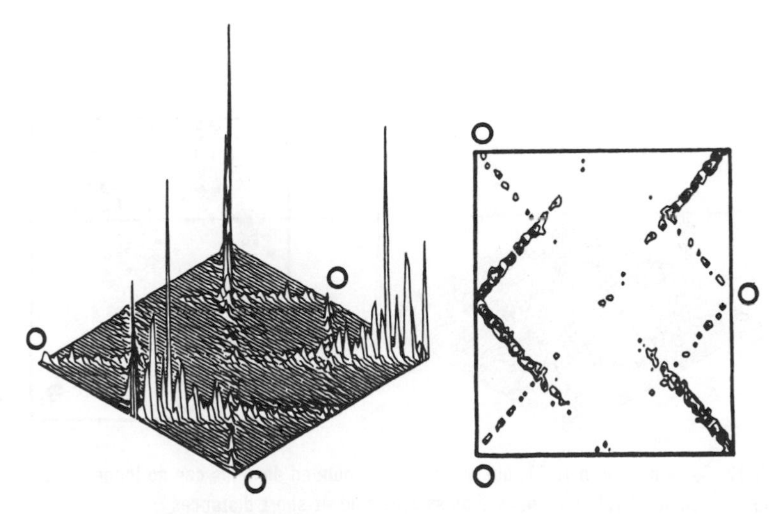

Figure 10.11 Two-dimensional configuration of a crystal of quaternane at 5 K showing bounded and orderly domains and the Fourier transform of this figure (G. S. Pawley)

temperature, tend to fracture, to dislocate and to scatter in the reciprocal space, even when the correlations are only valid at short distances in the direct space. The authors' astonishing images inevitably call to mind the formation of mountainous landscapes as set out by Voss and Mandelbrot.

This scattering of the boundaries turns up again in the diffraction spots obtained on the fractal interfaces constructed and published by Mandelbrot. The self-similarity in the reciprocal space is quite evident and can serve to guide the study of highly disordered materials at either the microscopic or galactic levels.

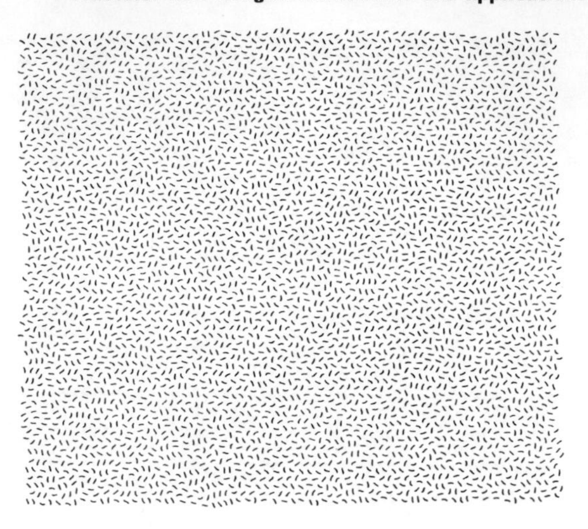

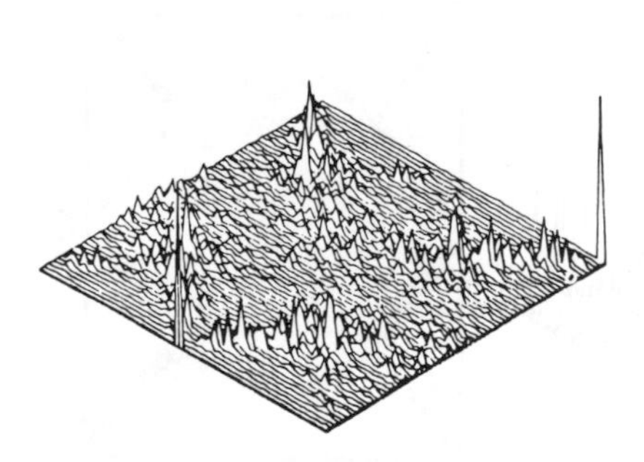

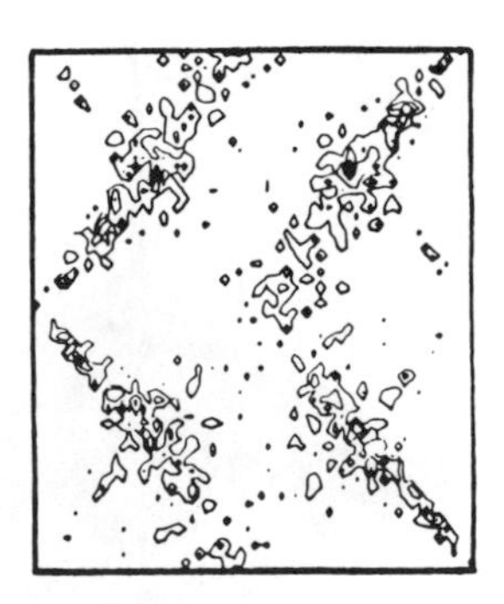

Figure 10.12 Same as Figure 10.11, but at 14 K. The bounded domains can no longer be seen by the naked eye. The basic correlations nevertheless still hold at short distances

10.6 Conclusion

We have just covered, in a few pages, a vast and as yet, for the most part, unexplored territory, namely that of the frontier regions between fractals and materials. How can the fractal concepts help us to develop new materials? Conversely, how can materials reveal fractal properties? These are the kind of questions that arise.

Chapter 11 Problems concerning the concept of fractal in electrochemistry

Michel Keddam
Laboratory of the Physics of Liquids and Electrochemistry, University P. and M. Curie

11.1 Foreword

The concept of non-integral dimension, originally developed in a purely mathematical context, has over the last few years seen itself fruitfully applied in a variety of disciplines. This quasi-explosive expansion is due in great measure to the works of B. Mandelbrot [12]. The chief attraction of the fractal concept lies in its universality which has the potential to make of it a particularly alluring unifying principle. The major obstacle encountered in the practical application of the fractal dimension to real physical objects arises from the extreme difficulty, if not the impossibility, of submitting them to the mathematical operations by which the fractal dimension is defined. Furthermore, they are only able, at best, to exhibit a non-integral dimension from a microscopic cutting. Lastly, they very often have a random structure which makes it very difficult to interpret the notion of internal similarity. These various points will be taken up in several of the contributions to this work.

In the realm of materials, one can, in a general sort of way, distinguish two types of implication of the fractal concept:

(a) The description of geometric properties of a material or of an object.
(b) The description of physical–chemical properties of this same material.

The main goal to be aimed at is then to connect the given results of the two approaches. A look at the large amount of material published in this area in recent years clearly shows that certain properties such as those dealing directly with the topography of interfaces (adsorption) can be linked to a physical approximation of the Hausdorff–Besicovitch material. On the other hand, much of the behaviour which sets off charge-transport phenomena points to other non-integral dimensions. Among these, the spectral dimension has been the object of some thorough research.

The notion of fractal has been copiously applied in the disciplines allied to electrochemistry. It is appropriate to mention the chemistry of solids [7] and, above all, the chemistry of interfaces [1]. The introduction of the notion of reaction on a fractal surface or of the percolation mass of reaction sites made way for an interpretation of a whole class of dynamical behaviours as yet unexplained although known for a long time and displaying a certain universal character. In the time domain, this had to do with laws of reduction deviating from exponential responses (the Curie–Von Schweib law). In the frequency domain, their duals exhibit well-known behaviour of complex susceptibility deviating from the lorentzian and interpreted by means of the empirical dispersion laws (Cole–Cole and its relatives) [14].

In electrochemistry, having regard to the predominance of measures in the frequency domain, it is not surprising that the concept of fractal was originally invoked in order to account for certain aspects of the data relating to measures of electrochemical impedance [8, 9, 11].

11.1.1 Some *a priori* considerations concerning the involvement of the fractal dimension in electrochemistry

The nature of electrochemical processes and their relationship with the science of materials allows us to identiy, *a priori*, the role of:

(a) A fractal geometry (of interphase frontiers) capable of describing roughness, porosity, sponginess dendritic interfaces, etc.
(b) A fractality of the spatial distribution of reaction sites: a surface and/or volume fractal network of sites.
(c) A combination of the above two types: fractal distribution of sites on an interface with fractal geometry.

A variety of corresponding situations is illustrated in Table 11.1.

11.1.2 Some comments on the TEISI model (transfer of energy on an interface with internal similarity)

To our knowledge, the TEISI model embodies the first attempt in electrochemistry to take account of an eventual non-integral dimension of the interface. It was proposed in order to attempt to explain a coherent group of experimental results indicating, quite unambiguously, that the frequency dispersion of the interfacial impedance just depends on geometric features of the interface. This fact is not new in itself but had previously given rise to models of the transmission-line type involving a large number of arbitrary parameters. The TEISI model had all the elegance of a description in fractal terms. All the same, it seems to us that there are a number of sticking-points in it, which we have tried to analyse below. Whatever the significance of these complaints, it must be stressed that the TEISI model represents important conceptual progress and a stimulant to research on the dynamic behaviour of the electrochemical interface. According to its authors, the

Table 11.1 Some ways in which non-integral dimension can be involved in electrochemistry

Geometric fractality of the interface		*Fractality of the spatial distribution of reaction sites*
Deviation from smoothness = roughness, porosity		
Static response		*Surface*
Effect of the size of the adsorbants		Fractality of islands of adsorbants
Adsorption isotherm		Fractality of germ-masses
Heterogeneous catalysis		Two- and three-dimensional surface phase transitions
Dynamic response:	time and frequency	*Volume*: fractality of the networks of reaction sites in a three-dimensional phase
Transfer		
Transport	TEISI model	Kinetics of volume increase of films
		Application of random walk models to site networks (surface or volume diffusion with the kinetic laws of electrochemistry)

value of this model lies, above all, in its practical effectiveness in the matter of the choice of materials for the electrodes of generators.

11.2 Theoretical considerations

The following comments are based essentially on the last published version of the model [8].

In its most elementary form, this model describes a transfer governed by linear laws on a fractal frontier. The fractal property is brought in by supposing that to each frequency in the sapce of Laplace frequencies (s) the effective surface on which the transfer of charge occurs, in the electro-chemical sense, is 'scanned' by a 'gauge' $\eta(s)$ in the sense of Richardson.

The physical existence of such a gauge can be justified when the physical process itself provides a characteristic length as in the case of diffusion at the microscopic level (Fick's laws) [3]. This involvement of the transport of matter is, however, formally set aside in the simplified model and called 'δ-transfer'.

As far as the TEISI model is concerned the form of η(s).

$$\eta(s) = \frac{\lambda_0}{s^{1/d}}$$

where d is the fractal (Hausdorff) dimension and λ_0 is an undefined characteristic length, is chosen in an *ad hoc* way to ensure agreement with experimental results [11]. The comparison is made in terms of electrochemical impedance, the complex current/potential relationship across the interface, and the model then comes up with an expression for a transfer function by using the principles underlying the modelling of dynamical systems (chemical reactors) in a linear regime (first-order reactions and linear transport).

There is nothing wrong with this in principle. Nevertheless, the model adopted clearly describes the dynamics of a return to equilibrium after a considerable thermodynamic perturbation at the frontiers of the system. The asymptotic behaviour for long periods of time or low frequencies (linked classically by the tauberian theorems of the Laplace transform [5]) corresponds to a zero flux (and therefore to a zero rate of creation of entropy). This linear dynamical behaviour can be represented by its electrical analogue in the form given in Figure 11.1. The force–flux (potential–current) impedance tends to infinity at low frequencies. The analogy with the formulation of the TEISI model in its degenerate form at a non-fractal interface is underlined by the choice of the same symbols for corresponding quantities. This dynamical model, which has recently been given the name 'blocking electrode' in electrochemistry is realized to perfection by films adsorbed by a reversible reaction, with modified electrodes and passive and electrochrome

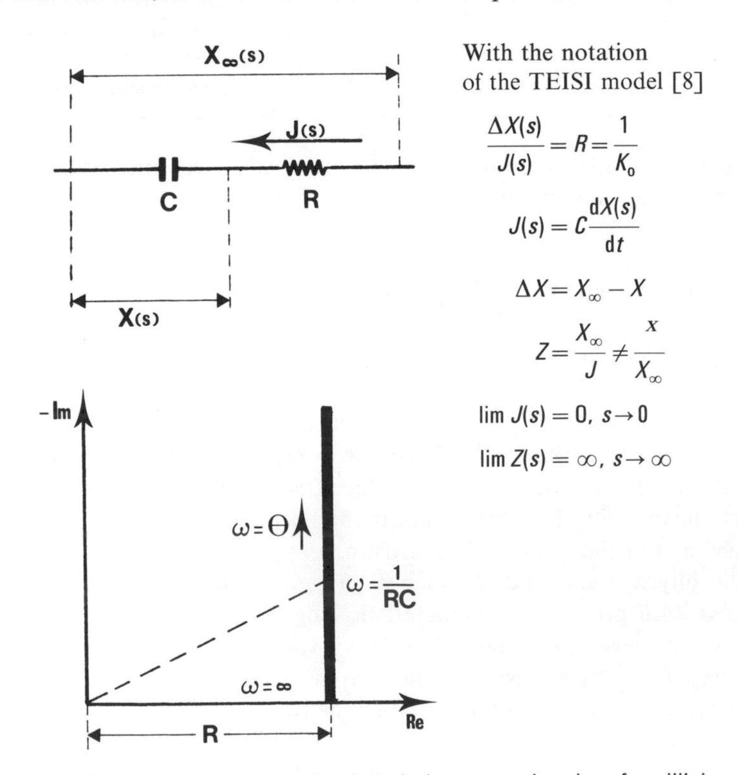

Figure 11.1 Electrical representation of a dynamical system: relaxation of equilibrium

films. More generally, it is valid for every system possessing an internal degree of freedom (the chemical potential of an adsorbed or condensed spot or one that is inserted), allowing it to settle to a new thermodynamical equilibrium following modification of a constraint at the frontiers (relaxation of equilibrium).

It is clear that this kinetic situation can only correspond to a system having a finite impedance at its bottom frequencies [10], such as that which the TEISI model is intended to deal with (Figure 11.2). For these interfaces, a perturbation of the constraints at the frontiers, even lying within the linear domain of TPI (the Onsager domain), creates a regime of permanent irreversibility. This model corresponds to a Red-Ox system for which even a mild boost ΔE applied to the potential of thermodynamic equilibrium creates a permanent flux $\Delta J = \Delta E/R$, where R is the charge-transfer resistance (relaxation of the permanent regime). In this last case, the capacity C is generally attributed to the electrochemical double layer and is often of an order of magnitude less than the capacity of Figure 11.1 (Faraday pseudo-capacity).

It really looks as if, in some sense, one is faced with some confusion, which is albeit very widespread in electrochemistry, between (thermodynamic) equilibrium and a state of permanent regime. The transfer function established in the TEISI model following comparison with impedances of the type of that of Figure 11.2 (in the case of fractal interfaces) is therefore unsuitable for this

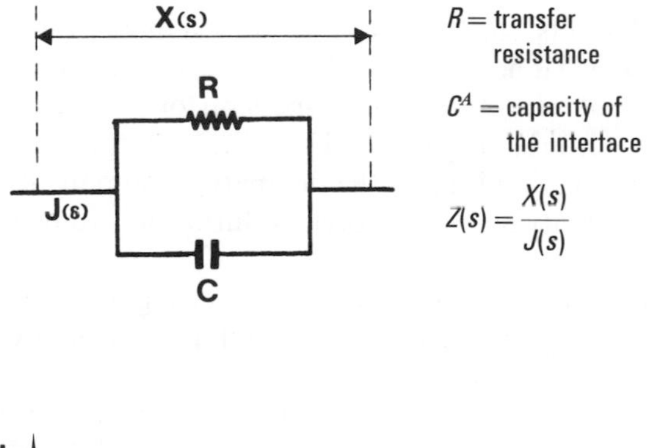

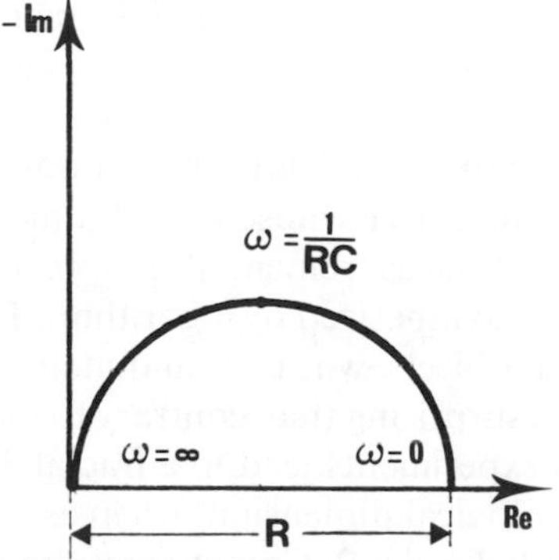

Figure 11.2 Electrical representation of a dynamical system: relaxation of permanent regime

purpose since it is based on a model of the type of those presented in Figure 11.1.

Appearances can be saved by choosing, instead of the (measurable) electrochemical transfer function $X^{\infty}(s)/J(s)$, the adimensional transfer function $X(s)/X^{\infty}(s)$ between two forces: one, $X(s)$, internal to the system and the other, $X^{\infty}(s)$, acting at its frontiers. Such a quantity is totally inaccessible in practice and is therefore of no interest.

11.2.1 Summary

To sum up, the TEISI model rests on:

(a) The hypothesis of a frequency-dependent exploration of the fractal interface by means of a (Richardson) gauge, the choice of which has yet to receive clear justification. In particular, it has not so far been possible to specify the frequency scale–spatial scale (the parameter λ_0) correspondence.
(b) A dynamic model of the interface which in the limiting case of a euclidean surface is not identified with the elementary electrochemical charge-transfer process that the TEISI model is meant to take into account.

11.3 Experimental considerations

The experimental results which underpin the TEISI model as well as the attempts at verification in which it has been used both raise a certain number of questions.

The large amount of data on the dispersion of interfacial impedances on smooth, rough or porous electrodes lend support to the proponents of the role of geometry and help to establish that the dissipative component of the dispersion depends on the resistivity of the electrolytic solution according to the predictions of the Kramers–Kronig relations [14].

These facts stimulate one, always within the setting of a representation of the interface by a non-integral dimension, to look for a model in which the specific dissipative character of the interface would be explicitly linked with the finite conductance of the electrolyte and not to the electronic transfer as in the TEISI model. This would therefore involve a direct extension of the de Levie model [4] to an interface with a particular metric. Two very recent attempts in this spirit may be mentioned [6, 13].

One, if not the only, way of putting these models to the test was by carrying out electrochemical measurements on an object whose fractal dimension(s) is (are) well known, either by independent measurement (for a real object) or by construction (in the case of an object synthesized by algorithm). The first of these two approaches, applied to natural objects, was chosen quite recently [2], with encouraging results. It is nontheless surprising that, contrary to the predictions of the TEISI model, electrochemical experiments lead to a fractal dimension lying between 1 and 2 for a surface (topological dimension) whereas the object under study has a fractal dimension d with $3 > d > 2$. Comparison should therefore be made between the electrochemical data and the fractal properties, not of the object itself but of cross-sections of it. It seems that this difficulty reveals certain 'inconsistencies' of the TEISI model which could possibly have their source in the points discussed above. It appears that the model based on the dissipative role of the resistance of the electrolyte does not come up against this contradiction [6].

References

1. Avnir D., Farin D. and Pfeiffer P. (1983) *J. Chem. Phys.*, **79**, 3566.
2. Clement E. (1984) D.E.A. Physique des liquides, Univ. P. and M. Curie.
3. de Gennes P. G. *CRAS II*, **295**, 1061.
4. de Levie R. (1967) In P. Delahay and C. W. Tobias (Eds.), *Advances in Electrochemistry and Electrochemical Engineering*, Interscience, New York.
5. Hladik J, (1969) *La Transformation de Laplace à Plusieurs Variables*, Masson Ed., Paris.
6. Keddam M. and Takenouti H. (in preparation) Communication submitted to the Congress of the Material Research Society, Boston, December 1985.
7. Klymko P. W. and Kopelman R. (1983) *J. Chem. Phys.*, **87**, 4565.
8. Le Méhauté A. (1984) *J. Statis. Phys.*, **36** (5–6), 665.
9. Le Méhauté A. and Crepy G. (1983) *Solid State Ionics*, **9–10**, 17.
10. Le Méthauré A. and Crepy G. *CRAS II*, **294**, 685.
11. Le Méhaute A., de Guilbert A., Delaye M. and Filippi C. (1982) *CRAS II*, **294**, 835.
12. Mandelbrot B. (1983) *The Fractal Geometry of Nature*, Freeman, San Francisco.
13. Nyikos L. and Pajkossy T. (to appear) *Electrochimica Acta*.
14. Scheider W. (1975) *J. Phys. Chem.*, **79**, 127.

Chapter 12 Some remarks concerning the structure of galactic clusters and Hubble's constant

P. Mills
Biorheology and Physico-chemical Hydrodynamics Laboratory, University of Paris VII

In chapter VI of this book *Fractal Objects* [1], B. Mandelbrot solves the problem of Olbers' so-called red sky paradox by devising a hierarchical structure of matter in the universe.

Consider an observer who measures the luminosity of the sky at some point of the universe. According to cosmological principles, this observer will come up with the same results whatever his place in the universe.

The luminosity of a star at distance R from an observer decreases as $1/R^2$, just like its apparent surface area. The apparent density of luminosity of the sky would consequently have to be the same for all stars and the sky would have to be uniformly luminous.

On the other hand, if the matter in the vicinity of the observer has a fractal structure characterized by the fractal dimension D, the mass–density, just like the apparent luminosity, decreases with R as R^{D-3}. Lastly, if $D < 2$, there are certain solid angles that do not contain any matter.

This fractal solution of Olbers' paradox is altogether both appealing and simple; nevertheless, it poses a few problems to do with Hubble's constant.

Consider a sphere of radius R centred at our observer and including a very large number of galactic clusters and let m be the mass of a galaxy at the surface of this sphere. The speed of the galaxy measured by the observer is given by Hubble's law:

$$V = HR$$

where H is Hubble's constant [4].

The total energy E of the galaxy calculated according to newtonian mechanics is the sum of the potential and kinetic energies:

$$E = mR^2[\tfrac{1}{2}H^2 - \tfrac{3}{4}\pi\rho(R)G]$$

where G is the gravitation constant and $\rho(R)$ is the cosmic mass–density. E must be zero for the galaxy to have exactly the escape velocity. The value of Hubble's constant is then obtained:

$$\tfrac{1}{2}H^2 = \tfrac{3}{4}\pi\rho G$$

It is still necessary for the critical density ρ of the cosmic mass to be independent of the spatial variables.

It inevitably turns out that the solution of Olbers' paradox give rise to a new problem. In order to solve the sky-colour problem, one supposes that the galactic density near our observer decreases as R^{D-3}, with $D < 2$, but one then discovers that Hubble's constant is no longer truly constant:

$$H^2 \sim R^{D-3}$$

The solution of these two paradoxes could be a compromise solution already proposed to describe clusters of particles in suspensions [2].

One considers a correlation distance ξ. In a volume ξ^3 the matter has a fractal dimension $D < 3$; ξ cannot tend to ∞ because the global density then tends to 0. At distances greater than ξ, the matter has a homogeneous structure. Let ϕ be the (volume) fraction of cosmic matter specified in a volume very much larger than ξ^3. We must have the relationship

$$\frac{\xi}{a} \sim \phi^{1/(D-3)}$$

where a is the characteristic dimension of the elementary dense particles* making up the fractal clusters.

The volume fraction ϕ can be deduced from the critical value of the universe ρ. As for the value of a, one can give it the mean value of the radius of a star in a galactic cluster.

Mandelbrot gives the value $D = 1.21$ as the fractal dimension of galactic clusters observed from the earth.

The structure of galactic clusters could be described qualitatively as follows. An observer, situated anywhere in the universe, would see compact connected aggregates in the foreground, then, at intermediate distances, connected fractal aggregates and then, at very great distances, scattered aggregates, separated by large voids. The mean density at these large distances could be regarded as constant.

Such a multifractal structure could be the result of competition between attractive interactive energies in the matter which tend to compactify it and the entropy terms which promote expansion of the fractal structures, and therefore of the weak dimensions, i.e. dimensions close to one for connected clusters [3].

*These 'particles' can be of the size of stars!

References

1. Mandelbrot B. (1984) *Les Objets Fractals*, 2nd ed., Flammarion.
2. Mills P. (1985) *J. Phys. Lett.*, L301–L309.
3. Mills P. and Snabre P. (in preparation) Equilibrium configuration of fractal aggregates.
4. Weinberg S. (1978) *Les Trois Premières Minutes de l'Univers*, Seuil Editions.

Chapter 13 Disorder, chance and fractals in biology

Guy Cherbit
Research Group in Membrane Biophysics, University of Paris VII

The literature of the subject is full of examples freely incorporating chaotic, random and fractal structure. When, further, these structures are spatiotemporal, the situation sometimes becomes somewhat complicated.

The examples that follow illustrate three types of statement currently to be found in biology.

13.1 Glomerular filtration (J. Schaeverbeke, M. Cheignon and S. Cornet)

Plasma filtration takes place at kidney level. It consists of functional units, the nephrons, which owe their filtration property to a specialized extracellular matrix, the glomerular basal membrane (GBM).

It intervenes in the blood filtration process simultaneously as both a molecular sieve and an electrostatic barrier, and will thus select the plasma proteins to be transferred to primitive urine as a function of their size and of their electric charge. This selective permeability is determined by the intrinsic chemical composition of the GBM.

The quality of the filter depends on the organization, interactions and purity of its constituents. These carry ionized groupings which can either be involved in electrostatic bondings or be free. Furthermore, it is these that carry the charge of the GBM.

These fixed negative sites can be shown up by means of a cationic marker of colloidal iron which sticks to them. By thus localizing them, it reveals the complexity of their configuration in space, which is not random (Figure 13.1). This visualization also enables one to follow the putting in place and the evolution of the distribution of these charges (and of the macromolecules that carry them) in the womb, during old age or in pathological situations.

The complexity of the membrane could be quantified by superimposing a grid on the plate and counting the colloidal grains. In log–log coordinates the slope of the straight line obtained would give the 'fractal dimension'.

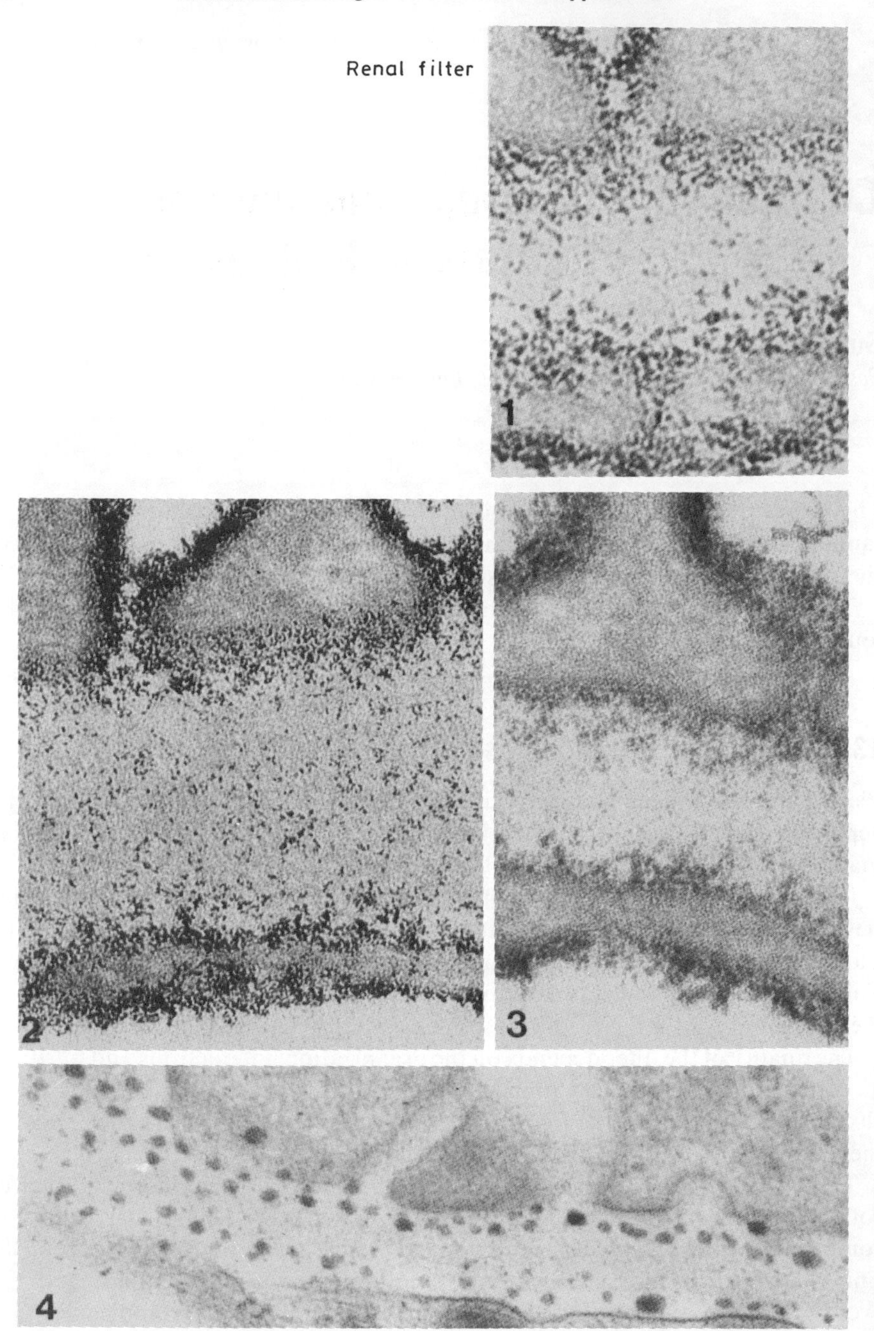

Figure 13.1 At the kidney stage, a very fine filter enables the blood to be purified. Cationic markers of colloidal iron, opaque to electrons, make it possible to show, using an electron microscope, the presence of negative charges at the core of the filter. The picture displays varying complexity according to the size of the grains of the marker. (*Plates*: M. Cheignon and S. Cornet)

Here, it is the size of the grains of iron colloid that provides the gauge of the measůrement, and the results are independent of the degree of enlargement of the plate as well as, and this is more unusual, of the magnification of the microscope.

One has an analogous representation by imagining a sort of comb whose teeth are coral-shaped and smeared with glue. Sprinkling this with sand, small polystyrene particles or ping-pong balls, one would obtain a 'pseudo-adherence' characteristic of the rake but dependent on the diameter of the covering grains (D. Avnir and P. Pfeifer).

13.2 Perturbation of the rate of growth of a culture (G. Cherbit)

To study the growth of a hyphomycete fungus in a Petri dish, it is enough, having sown the monospore, to measure the diameter of the culture as time passes.

In view of the fact that the growth is the outcome of a contest between the speed of implantation and the speed of penetration, one obtains the radius of the culture afte time t:

$$R = \lambda(1 - e^{-(\mu t)^{3/2}})$$

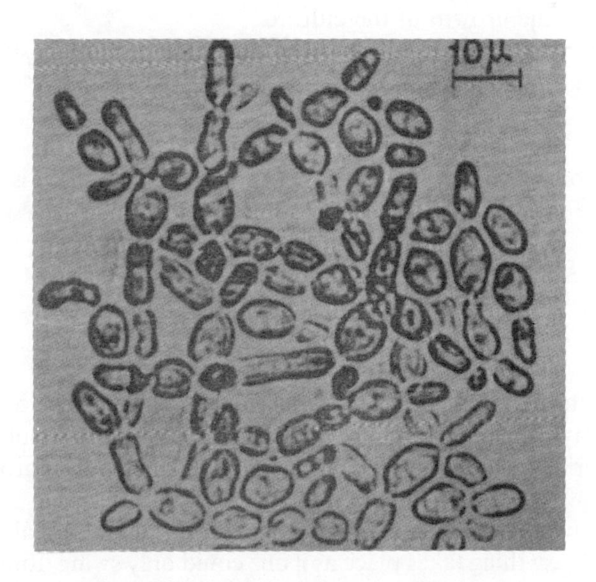

Figure 13.2 Monospore culture (48 hours after sowing). The number of branches is proportional to the radius of the circle which contains them, thus verifying the random hypothesis. (*Plate*: Caporali and G. Cherbit)

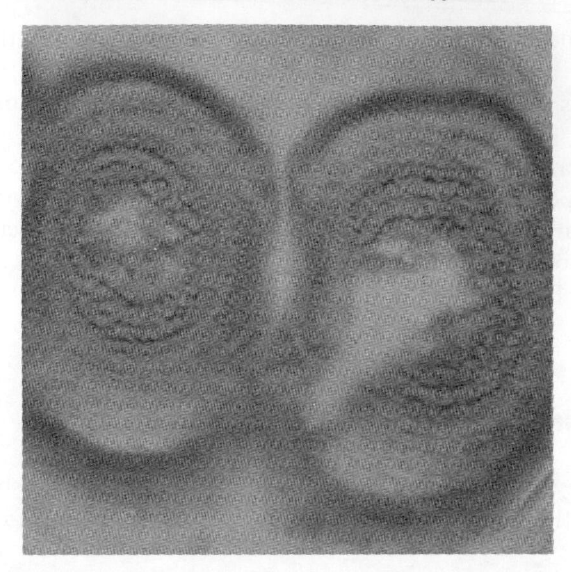

Figure 13.3 Culture of two colonies of Metarrhizium An. showing the competition between penetration and implantation. (*Plate*: G. Cherbit)

where λ denotes the radius of the box and μ is a parameter that depends on the transmembrane exchanges of the fungus (Figure 13.2 and 13.3).

Starting with this approximate model, one can study the effect of a perturbation on the growth of the culture.

Consider then a pulse-type perturbation (light pulses, thermal shocks, etc.) and its encoding in binary form:

$$1\ 0\ 0\ 1\ 1\ 0\ldots, \text{ for example}$$

(1 for pulse, 0 for the absence of a pulse). A curious phenomenon is then revealed: if the sequence is periodic

$$1\ 1\ 0\ 0\ 1\ 1\ 0\ 0\ \ldots, \text{for example}$$

or if the perturbation is continuous

$$1\ 1\ 1\cdots 1\ 0\ 0\ 0\cdots 0 \quad \text{or} \quad 0\ 0\ 0\cdots 0\ 1\ 1\ 1\cdots 1$$

then the growth of the culture slows down (for the same energy of perturbation, i.e. for an always constant value for the sum of the coding numbers).

On the other hand, if the sequence is random, then the rate of growth increases (Figure 13.4).

This experimental result is borne out, whatever the physical nature of the perturbation. Everything takes place as if one could only swing from one attractor to another by means of a random perturbation.

Such a situation is seen very often. Nevertheless, it has not yet received a satisfactory explanation, although the work of the Brussels School ought to provide a usable formalism in this context.

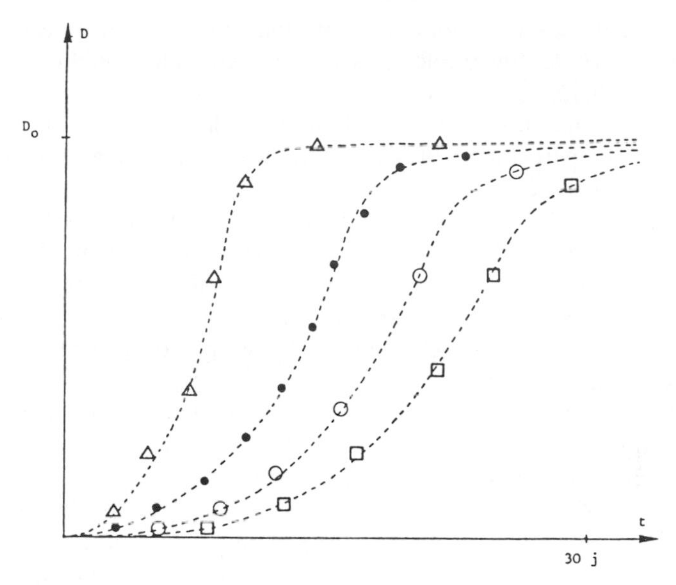

Figure 13.4 Perturbed growth curves. Random perturbations increase the rate of growth
△ Random perturbation □ Continuous perturbation
○ Periodic perturbation ● Unperturbed

13.3 Bacterial Division (Marcovitch)

Mitotic division of bacteria takes place according to the cycle shown in Figure 13.5. In a culture, the proportion of bacteria at different stages is constant, both in time and from one part of the culture to another.

It this dual property that commands our attention. It is easy to pass, by ergodicity, from the time distribution of the different stages to their spatial distribution.

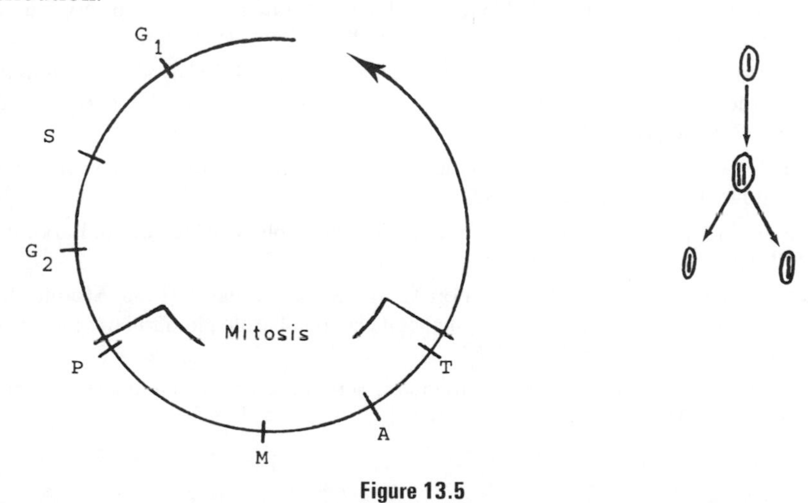

Figure 13.5

From Mandelbrot's first work, we note that, in a fractal object, 'a part is identical to the whole' (invariance of scale). Marcovitch thought that this was strictly the case here.

Nevertheless, it has not been possible to 'fractalize' this situation. (Did this really matter? The percentages of bacteria at the different stages provide a good 'quantification' of the culture).

Questioned about this example, Mandelbrot replied that a fractal is above all a *geometric object* and that the problem posed by Marcovitch did not have a fractal solution, reviving Plato's 'Nothing is admitted here if it is not geometry'. If it is true that fractal objects constitute a special case, then certain applications (several developed by Mandelbrot himself) do indeed describe non-geometric situations: processes, econometrics, etc.

In the present volume, the article of M. Weber (Chapter 5) provides tools for such applications.

Of course, one can often geometrize a situation, but this is not accomplished without some modification. For example, the study of the motion of a particle cannot be reduced to that of its trajectory. Thus, if one projects a brownian motion onto a line, one obtains a segment of this line which is the trajectory of the projected point but does not show the complexity of the motion of this point.

The problem posed by Marcovitch is therefore important for the deep questions that it raises and also because it illustrates a situation encountered each time one wishes to study the kinetics of a non-temporal random distribution which is also non-geometric.

Bibliography

Avnir D. and Pfeifer P. (1983) Fractal dimension in chemistry. An intensive characteristic of surface irregularity, *Nouveau Journal de Chimie*, **7**(2), 71–72.

Avnir D., Farin D. and Pfeifer P. (1983) Chemistry in non-integer dimension between 2 and 3, 2: Fractal surfaces of adsorbents, *J. Chem. Phys.*, **79**(7), 3558–3565.

Cheignon M., Bakala H., Geloso-Meyer A. and Schaeverbeke J. (1984) Modifications de la barrière de filtration glomérulaire au cours du vieillissement chez le rat, *C.R. Acad. Sci., Paris*, **299**, Série III, no. 9.

Cherbit G. (1976) Modèle mathématique des croissances myceliennes sur milieu synthètique, *Cybernetica*, **19**, 75–82.

Cherbit G. (1986) Chaos et entropie, Séminaire Interdisciplinaire Hausdorff, Laboratoire TMIB, University of Paris VII.

Cornet S., Chami S., Bakala H., Cheignon M. and Schaeverbeke J. (1986) Modifications structurales, biochimiques et fonctionnelles de la lame basale glomérulaire au cours de la senescence, *Biology of the Cell*, **57**.

Pfeifer P. (1986) Catalyse et surfaces fractales, Séminaire Interdisciplinaire Hausdorff, Laboratoire TMIB, University of Paris VII.

Chapter 14 Fractals, semi-fractals and biometry

Jean-Paul Rigaut
Research Unit: Biomathematics and Biostatistics, University of Paris VII

—but who is going to examine the coast of Brittany under the microscope?

14.1 Introduction

There is no lack of examples, in Mandelbrot's inspired work (1982), of natural objects, at least part of whose form lends itself to the setting of a fractal model. The assertion of the author ('there is a fractal side to nature'), although abundantly supported by a wealth of striking facts and most convincing explanations, nevertheless leaves a fundamental ambiguity hanging in the air: *is* nature really fractal, or does fractal geometry merely offer an, albeit marvellous, tool for describing it? In the first case, explanatory questions should be asked: how can the *structures* encountered in nature result from a construction based on fractal principles? In the second case, it is still impossible to avoid the equally explanatory question of 'why'? The usual fractal models are basically too 'perfect' to give a totally accurate explanation of the origins of a natural structure, other than in a purely geometric way. The introduction of random elements clearly increases the possibilities. Could it be that the observed result, the outcome of a multiplicity of probabilistic phenomena, is *close* to 'fractality', which itself represents an *ideal*?

It has seemed to us, in the course of our own work, that nature *tends* to fractality. It is thus possible, as we shall see, to avoid the somewhat unnatural problem of the existence, in an ideal fractal model, of lengths tending to infinity. It should be said right away that this is bought at the expense of the relative ease of comprehension that ideal fractal models offer; the existence of a *unique* (or of several) non-integral dimension(s), which can be associated with an object, will generally be replaced by a 'continuous derivative of the fractal dimension', with 'ideal' fractal asymptotic properties. The corresponding objects will be called 'semi-fractals'.

It will appear, we hope, that this new approach offers a setting in which the dilemma of Mandelbrot's assertion can be avoided. Let us try a slogan of our own: 'natural constructions are semi-fractal and tend to a fractal ideal'. Very often, we believe that this ideal can quite easily be identified: this is the result, or the final product, to which a process tends. It can be investigated by means of semi-fractal models and is certainly of a probabilistic character.

It is reasonable to ask if the study of nature by means of ideal fractal model camouflages deeper laws. Explanatory models are essentially superior to the purely descriptive ones often presented in the fractal context. May we tentatively hope that out modest contribution will provide food for thought in this direction? We think that to give more precise descriptions is the very least that is required in eventually arriving at explanations. At first sight, certain objects can be regarded as ideally fractal, although this is often just an approximation which nevertheless turns out to be a powerful tool in practice. Our semi-fractal model gets closer to the true nature of the natural objects that we have looked at. We shall see that it is very useful in practice and yields new information.

Mandelbrot (1982) has raised the possibility of a 'slowly drifting' fractal dimension in a short and, alas, isolated paragraph. It is this to which we have been irresistibly drawn. If we may paraphrase Cantor (in Mandelbrot, 1982), it should be said, after having analysed our experimental findings, that, for us, 'to see was not to believe'! We were not able to use the Richardson–Mandelbrot model for the boundaries of our objects, and we did not wish, like some authors, stubbornly to adjust the graphs of log–log curves using a series of straight lines in order to cling to the usual fractal theory. It is not so difficult to reason in the setting of a drifting fractal dimension. In contrast with Mandelbrot, we shall not regard this as a special case, but, on the contrary, as the most general that there is. We would not like their being called 'semi-fractal' to cause our objects to be treated with contempt; they are not marginal; may be they are the 'true' fractals, the others (the pure ones) merely their idealized form?

Nature has a fractal side? Well, so be it. Let us look at this *closely*. Coastlines, mines, etc., are looked at 'from afar'. In our world, biometry, the microscope rules. All the same, we believe that our results have a more general value—but who is going to look at the coast of Brittany through a microscope?

14.2 History

Only a few works in biometry have made use of fractal geometry. We will mainly be concerned with the biometry of tissues examined under the microscope. All the same, we should mention several other types of work in biology. It is tempting to evaluate the fractal dimensions of plants; everybody knows about the more or less homothetic aspect of their characteristic branching structure (Mandelbrot, 1982); allometric relations of the type $y = ax^{\beta}$ have been described in phyllotaxy (Jean, 1983); the study of the fractal characteristics of trees can turn out to be useful in practice, notably in ecology (Loehle, 1983; Morse *et al.*, 1985); the forms of flowers have been simulated unbelievably well by computer using a fractal model (Mandelbrot, 1982). Furthermore, the convolutions of the brain and the physiology of the capillary circulation of blood can best be understood by means of fractal geometry (Mandelbrot, 1982).

In microscopic biometry, in our case as well as in those of the other authors that we shall look at, the surfaces of biological tissues are what, for the most part, form the object of interest (i.e. the outer boundaries of tissues or membranous surfaces). As the experimentalist generally begins work with histological cuts, he or she measures lengths: those of the cross-sections of surfaces in the thickness of the microscopic cut. The areas of these surfaces are estimated on the basis of these lengths, with the help of stereological formulae (Weibel, 1979).

A certain unease began to be evident when certain workers noticed that the stereological estimates varied from one author to another, for the same biological object. Reith (1976), Reith *et al.* (1976) and Bolender *et al.* (1978) noticed that the area density by volume (S_V) of the surfaces of subcellular membranes was estimated in a wildly different way from one work to another: 5.7 μm^{-1} for the hepatic endothelial reticulum of the rat according to Loud (1968), as against 14.1 μm^{-1} for Weibel *et al.* (1969); 9.7 μm^{-1} for the internal membranes of mitochondria for Reith (1973), against 32.5 μm^{-1} for Weibel *et al.* (1969). Bignon and André-Bougaran (1969) had been the first to note that such differences could be attributed to different resolutions (i.e. most frequently, to different magnifications of the microscope), in their case, for the area of the pulmonary alveolar surface. Keller *et al.* (1976) were the first to study the effect of the resolution systematically; they found that the value of S_V for human pulmonary alveoles varied from 180 to 300 cm^{-1} (180–240 in optical microscopy) according to the magnification at which the measurements of the alveolar boundary were carried out. This explains how the total area of these alveoles could have been estimated at between 80 (optical microscopy) and 140 m^2 (electron microscopy) in man, depending on the author (Weibel, 1963; Thurlbeck, 1967; Gehr *et al.*, 1978). Similar effects have been described for the morphometry of bones (Olah, 1980).

It should be noted that the effect of the resolution of the microscope had previously been observed in the science of materials (Underwood, 1961) and that a fractal interpretation had been proposed by Coster and Deschanvres in 1978.

In biology, Weibel (1979), following his meeting with Mandelbrot, announced his estimate of a fractal dimension (on a log–log graph) of 2.17 for the alveolar surface of human lungs (see also Mandelbrot, 1982). However, the first really systematic work using a fractal model in biology was published by Paumgartner *et al.* (Weibel's group) in 1981, after having been presented at a stereology Congress at Salzburg in 1979. These authors studied, on electron micrographs, cuttings from subcellular membranes of rat hepatocytes. The S_V values were estimated by the classical stereological technique using the intercepts with grids of lines on transparent sheets (cf. tracing paper) (Weibel, 1979), at magnifications ranging from $\times$ 18 000 to $\times$ 172 000. The estimates of the S_V thus ranged from 2.4 to 10.4 μm^{-1} (per unit volume of cytoplasm) for the endoplasmic reticulum and from 8.4 to 23.6 μm^{-1} (per unit volume of mitochondria) for the internal membranes of mitochondria! Even the volume density (V_V) varied (from 0.5 to 0.15 for the reticulum). The authors discovered a linear log–log relationship between the lengths (B) of the traces of sectioned membranes ($B = (\pi/4)AS_V$); with the method of intercepts, $S_V = 2I_L$, i.e. twice the number of intercepts between the lines and the membrane traces, per unit length of lines projected on the micrograph (Weibel, 1979) and the resolution. They therefore used the empirical model of Richardson (1961), interpreted in the fractal setting by Mandelbrot (1967), to estimate the fractal dimensions of their membranes. They estimated B at several magnifications, instead, like Richardson (1961) on his coastlines and frontiers, of using polygonation of the traces by a series of equal steps; as we shall see, this comes to the same thing (the step-scale takes the place of 'resolution'). Their log B versus log (resolution) graphs were acceptably linear except for values obtained with the strongest magnification (× 172 000), where no increase in B compared with the next most powerful magnification (× 130 000) could be detected. For an 'ideal' fractal structure, B should continue to increase (to infinity); also, following Mandelbrot (1977), the authors concluded that, for natural objects, a limit should be reached, beyond which 'the dimension becomes a topological dimension ($D = 2$ for surfaces)'. The magnification of × 130 000 would thus correspond, for these authors, to a 'critical resolution, above which the entire complexity of the membrane cross-sections is clearly observed'. Lastly, the authors proposed to use the S_V extrapolated to the critical resolution in order to make comparisons between different works using different magnifications.

Thus, around 1980, certain facts began to be properly formulated: the need to take account of the resolution was confirmed by all work in morphometry and the non-passage to the limit of boundaries of the objects was recognized; however, the log–log graphs were regarded as linear, allowing an estimate of a fractal dimension to be made, at least within a certain range of resolutions (magnifications). Thhe consequences of such facts for stereology were recognized by Weibel (1979) as possibly 'serious'. Moreover, this author explained quite brilliantly, taking the lung as an example, that regarding the alveolar surface as strictly self-similar would not be very satisfying for the physiologist interested in exchanges across the alveolar cells, but that log–log graphs with broken lines (corresponding to several successive fractal dimensions) would be more realistic from this point of view.

Several other authors have been interested in the fractal geometry of the lung. Barenblatt and Monin (1983) used homothetic principles to compare respiratory exchanges in pelagic animals; they suggest a possible fractal explanation. Mandelbrot (1982) briefly dealt with the geometry of bronchial division. Lefèvre and his collaborators (Lefèvre, 1983; Lefèvre and Barreto, 1983; Lefèvre *et al.* 1982, 1983) have proposed a remarkable model of the pulmonary vascular layer, based on fractal trees and a cost function. Both symmetric and asymmetric tree models have been looked at. The authors define 'solid' trees following Fredberg and Hoenig (1978) and using the Horsfield index (Singhal *et al.* 1973). Since the value of this index is $IH_s(n) = n + 1$, where n is the level of the branches, for a firm symmetric tree the simplest asymmetric firm tree has index $IH_{as}(n) = n + 2$; this corresponds to a Fibonacci tree (the class of such trees is defined by $IH_{f,k}(n) = n + k$). Perhaps the reader will in due course detect some kinship with our class of semi-fractal Fibonacci monsters?

Finally, Mandelbrot (1982) uttered the aphorism according to which 'the Lebesgue–Osgood monsters are of the same stuff as our flesh', thinking of capillary networks, although it seems that no experimental work has tried such an approach.

The small amount of work in biometry using the concepts of fractal geometry is really rather disappointing, but we are willing to bet that this state of affairs will not last long... unless the 'dawn of new understanding among biologists' calls for 'perhaps a generation'? (Thom, in Zeeman, 1977).

14.3 Experimental methods

14.3.1 What are we looking for?

In biometry, and especially with the microscope, the appearance of objects' shapes is the main source of interest as far as fractal geometry is concerned. A research area of some priority is the search for a way of quantifying the boundaries of biological objects. This is also true of much work in materials science, where fractures are also of supplementary interest, and in geostatics, and the methods that we are going to describe have almost always been tried in these areas before being applied in biology.

Fractal geometry provides a new tool: the fractal dimension. We shall not get involved in discussions about its definition again here. What is interesting for the biologist is that it expresses a degree of irregularity of the bidimensional outline or of the external surface of the object being studied.

Among the available methods for the determination of the fractal dimension of an object's outline, the most plentiful are those which involve the variation of the length of the outline as a function of the resolution. They vary in the way changes of resolution are obtained.

14.3.2 Variation of the perimeter as a function of the resolution

Method of variable magnification

This is the method used by Paumgartner *et al.* (1981): the perimeter is always measured in the same way, but on images taken at different magnifications (see above). The principle is simple: increased magnification, giving increased resolution, makes visible smaller and smaller irregularities. Intuitively, there is an equivalence between this resolution in the strict sense and the effects of type resolution obtained by other techniques that we shall see, but the analogies are not easy to formalize in most cases.

Equal step polygonation

This is the method that Richardson used (1961) and which enabled him to discover his empirical law on the length of coastlines and frontiers between countries. A series of equal straight line segments (steps) is fitted to the curve under study and this polygonation is repeated, varying the length of the steps. The length of the curve for a given step is then given by

$$B_r = N_r r$$

where N_r is the number of steps of length r. If, at the end of such an exercise, one comes across a fragments of curve too small to accommodate a step, the usual procedure, suggested by Richardson (1961), consists of adding to B_r the length of the straight line joining the two ends of this fragment. The effect of this device has never, to our knowledge, been studied systematically, but it would seem to be negligible in general. Another effect difficult to evaluate is that associated with the choice of starting-point for the journey along the curve.

This is the method that we have used most often, and is at the root of an algorithm for automatized image analysis.

Mandelbrot's interpretation (1967) of Richardson's law has been one of the building-blocks of fractal theory. If the graph of log B_r against log r is linear, then

$$B_r = \lambda r^{1-D}$$

where $1 - D$ is the slope of the graph. Evidently

$$\lambda = N_r r^D$$

and λ is a constant measure, in the dimension D, of B_r: for example, if $D = 1$, we are dealing with a normal length, but if $1 < D < 2$, the object has a fractal shape which becomes more convoluted as D increases.

The sausage method

This refers to the 'Minkowski sausage', here obtained by expanding a curve by means of a circle of radius r, whose centre is placed at a succession of points of the curve, sufficiently close to one another. Curiously, this method, investigated by Flook (1978, 1982), takes up an idea of Cantor (see Mandelbrot, 1982). The operation is carried out with the help of an image analyser; since it is not possible to put a true circle into the image memory of such an instrument (digitized image), one expands the curve by a 'structuring element' (the expansion/shrinkage operations are part of the stock-in-trade of computer image-analysis, and are formalized in the setting of mathematical morphology; see, for example, Serra, 1972) for the most isotropic possibilities (hexagon for a hexagonal frame or octagon for a square frame). The area of the sausage, easy to measure with an image analyser, divided by its breadth ($2r$), gives an estimate of B_r, which turns out in practice to be very close to that obtained by polygonation with steps of length r. One can expand in successive stages to obtain estimates of B_r for variable r. The problem of overestimating due to expanded extremities of curves with free ends (rare in biology) has been dealt with by Flook (1982), who introduced a correction factor.

Method of 'censured' intercepts

Due also to Flook (1982), this is based on the stereological formula for estimating B starting from intercepts (see above). Here, these are chords, i.e. intercepts of the object with horizontal lines of pixels in the memory of the image analyser. The 'resolution' is represented here by the minimal length of chord that can be measured for a contour which recrosses a test-line. The image analyser can even be calibrated at the outset to take account of only those chords exceeding a certain specified length.

Polygonation with widely spaced vertices of a number of successive pixels

This method, again created for the image analyser (Schwarz and Exner, 1980), puts together a polygonation of a curve by taking as vertices pixels that are n pixels apart all along the curve. It is difficult to assess the distorting effect of this, compared with taking true scale-steps. There is a risk of its being considerable for certain very tightly twisted curves. The choice of starting-point for the journey along the curve runs the risk of having more serious consequences than for the method of equal steps. The algorithm, however, is speedier.

14.3.3 Methods based on other principles

The area–perimeter relationship

Mandelbrot (1982) proposed a method for determining the fractal dimension (D) based on the study of the relationship between the perimeter and the area, with the same resolution, of sections of an object, or of realizations of varied sizes of an object of the same type. Fractal theory supplied the relationship $A \approx B^{2/D}$, which had been endorsed experimentally, for example, for clouds (Lovejoy, 1982) and the cut surfaces of metal fractures (Mandelbrot *et al.*, 1984). Nothing of this kind seems to have been tried in biology. Let us just note that Mandelbrot (1982) puts forward a relationship between superficial area and volume for the brain, with $D \approx 2.7$, this rather high value closely reflecting the very convoluted nature of the surface of the brain, which spreads out to fill the corresponding volume.

The number–resolution relationship

Korčak's empirical law on the number of dots observable at each level of resolution, in its fractal interpretation (Mandelbrot, 1982), enables one to estimate D. One counts the number of dots seen at each resolution, i.e. those having a diameter greater than a limiting value ∂, just detectable at this resolution. Then $N(\Delta > \partial) = K\partial^{-D}$. This could perhaps be applied in stereology, when one studies sections (representing the 'dots') of three-dimensional objects.

Variogramme and fractal dimension

Mandelbrot and Wallis (1969) have shown that the family of linear brownian motions (Bachelier–Wiener–Lévy processes) can be used to model the stochastic properties of geophysical phenomena. The variogramme of the increments (h) of the corresponding functions is of the form $|h|^{\beta}$, where $\beta = 4 - 2D$ (Mandelbrot, 1982; Burrough, 1984). The Weierstrass–Mandelbrot function (Berry and Lewis, 1980) has been thus used as a model for estimating D in geostatics (Burrough, 1981) based on the variogramme-type data available in the literature. Later on, in connection with our experiments on the variance of dispersion of the area-density, we shall again meet the variogramme (Matheeron, 1965) and we shall then see its definition.

Fourier analysis and fractal dimension

A Fourier analysis of the outlines of objects enables D to be determined (Mandelbrot *et al.*, 1984): the integrated spectrum takes the form K^{-B}, where $B = 6 - 2D$.

14.3.4 Our method for studying the outlines of objects

We use an automatic image analyser to polygonate the outlines of objects digitized using a television camera and then binarized. The method of scale-steps is applied by travelling along the outline automatically, looking, at each pixel, for the next pixel, which should be at the closest possible distance to the length of the step (r). The procedures for different steps are now carried out together; the B_r are memorized, as well as, for each step-length, the number of points of inflexion (DeHoff, 1981) and of positive and negative horizontal tangents to the objects (Cahn, 1967; DeHoff, 1967), because these quantities are necessary for other stereological calculations, as we shall see later on. The perimeter B_r for a given r (we consider steps of from 1 to 100 pixels) is the sum of all the consecutive distances for this step (each approximately equal to r) plus, when there is a 'leftover fragment', the distance between its two extremities. Our algorithm (Rigaut *et al.*, 1983b), completely automatic, is written in FORTRAN-80, and calls on matrix-processing microprogrammes in ASSEMBLER for the IBAS (Kontron, R. F. A.). This last is a square-framed instrument with which images in 256 shades of grey and 512 × 512 pixels can be memorized. In a preliminary version (Rigault, 1984), the different step-lengths were handled separately and the actual steps were not memorized, so that the theoretical steps ought to be corrected, to take account of the square frame (after having confirmed the absence of anisotropy).

We also used the method of dilation of contours, but this was not well suited to one of our favourite experimental models, the lung, as it causes the two edges of the contours of the interalveolar septa to join up very rapidly. Finally, we shall see that we used the variogramme in a study that seems to have promising connections with fractal geometry.

14.4 A Semi-fractal model

14.4.1 Origins

The development of our model was motivated by a simple observation: our ('Richardson–Mandelbrot') log–log graphs of B_r (the perimeter of an object) against r (step-length) were almost never linear (Rigaut, 1984). Most often, they were clearly concave throughout the range of resolutions investigated (see later, and Figure 14.1); sometimes, a part of the graph was acceptably linear, always for the greatest values of r (weak resolutions).

We only came across entirely linear graphs for the coastline of Great Britain (*Oxford Atlas*, 1:2 500 000, coastline traced with black felt-trip pen on tracing paper; $D = 1.28$) and for a speck of mineral dust with the optical microscope (at several magnifications, from 0.91 to 4.32 pixels/μm, always with different step-lengths at each magnification; $D = 1.05$).

The graphs were concave for all the other objects studied: by optical microscope: lung sections (distended or not—see next section), nuclear contours of erythrocytes and human blood lymphocytes and the direct image of the decagonal diaphragm of Köhler from the microscope with a level of grey insufficient for binarization (making the diaphragm look irregular); micrographs ($\times$ 14 450) from electron microscopy; the cellular outline of a foetal blood erythroblast of a rat and the nuclear outline of a human pulmonary macrophage; macroscopic objects: the young leaf of *Geranium rotundifolium* L., at several magnifications (46–455 pixels/cm), a black circle hand-painted 'as precisely as possible' and a round metallic chronometer with three push-buttons.

This 'list of missing things' is intentionally somewhat diverse in order to show clearly that most of them yield concave graphs. This concavity is quite pronounced in many cases. To estimate a dimension D from the slope of the graph ($= 1 - D$) thus seemd unsatisfactory for the majority of the objects studied. Note that the mineral particle was possibly not fractal in character (D very close to 1) and that, apart from this, the only linear graph was obtained from an object (a coastline, cf. Richardson, 1961) 'viewed from afar', i.e. *at very weak resolution relative to its detailed structure*. Furthermore, we repeat, our concave graphs sometimes exhibited a linear part at weak resolution. We shall see that this assumes a special significance in the setting of our model.

Figure 14.1 shows the log–log graphs that we obtained for the outlines of microscopic sections of the pulmonary alveoles of prematurely born rabbits, badly developed (untreated) or well developed (treated at birth with surfactant). They were obtained by means of polygonations with steps of from 1 to 100 pixels (2.3 pixels/μm). Note the concavity (particularly for the underexpanded lungs) and the tendency to linearity at weak resolutions (large step-lengths).

We have combed the literature—not that of biology, which is something of a desert from the fractal point of view, but mainly that of material science—to see whether other authors have indicated the existence of concave graphs. We were in

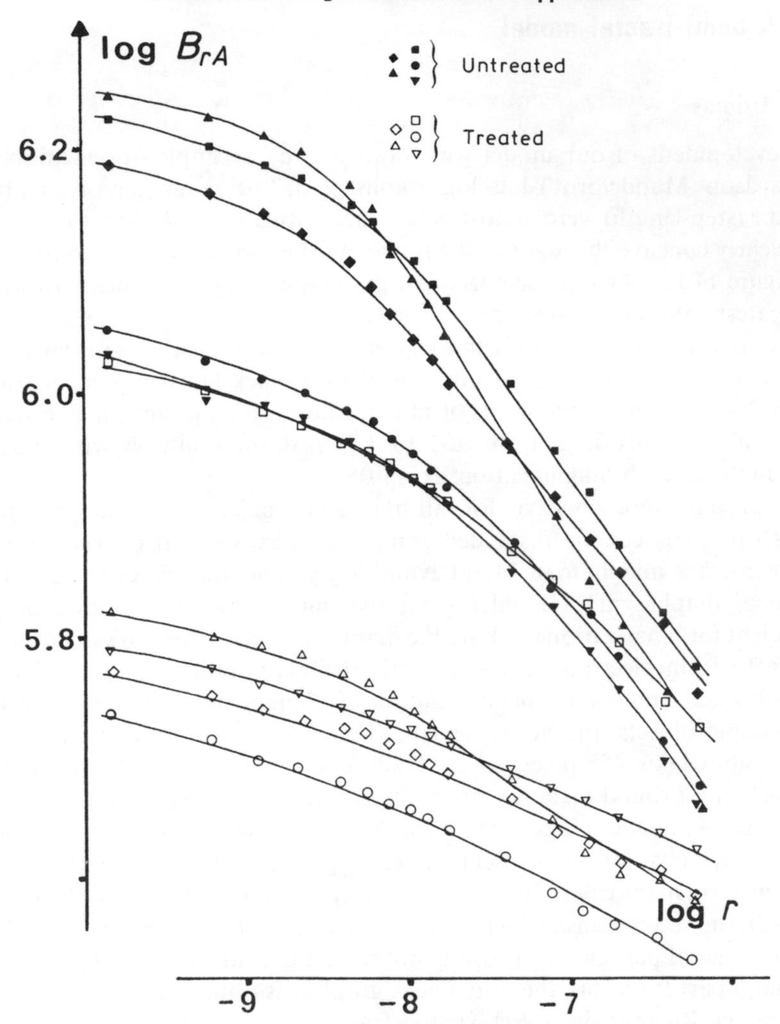

Figure 14.1 Results obtained on lungs; see Table 14.1 for the experimental conditions (only the 60% groups are represented here). The Richardson–Mandelbrot log–log graph: density of alveolar perimeter per reference area as a function of step-length. The results do not display the linearity which indicates an 'ideal' fractal, although such behaviour is evident for weak resolutions (large step-lengths), particularly for the treated animals (well-developed alveoles). Note that the curve for one of the treated cases is situated amongst those for the untreated ones (underdeveloped alveoles), although its form is different; compare Figure 14.3

for a double surprise: yes, examples of such graphs were sufficiently plentiful to reassure us—we were not alone (Coster and Deschanvres, 1978; Flook, 1978, 1979, 1984; Kaye, 1978; Schwarz and Exner, 1980; Orford and Whalley, 1983)—but their interpretation was generally different from that which we would have given from our own analysis!

Most of these authors fitted several linear segments to perfectly good curves. Perhaps this is due to a wish to 'cling' to the classical fractal model: a different dimension for each segment would indicate passage from one type of structure to another (Mandelbrot, 1982) or from a 'structural fractal' to a 'textural fractal' for the same object, then called 'multifractal' (Kaye, 1977). Only Orford and Whalley (1983) have expressed reservations: 'although this possibly goes against the spirit of fractal self-similarity proposed by Mandelbrot (1982), particles can very well exhibit continuous gradients of fractal change'. However, these authors, even in the same article, fit two to three linear segments to a curve, and this after having stated that no investigation had been undertaken to see if these interpretations were of any statistical value! Finally, certain authors (Chermant and Coster, 1978; Chermant *et al.*, 1981) have tried to attach some structural significance to the faults and irregularities of their graphs (metal fractures), which are more properly attributed to 'noise', of no real value, according to Wright and Karlsson (1983) and Mandelbrot *et al.* (1984) We should say that, in biology, the only graphs other than our own, namely those of Paumgartner *et al.* (1981), regarded by these authors as linear with a fracture at high magnifications, could equally well be viewed as concave: our equation (see later) accommodates all their values to a curve, with a correlation coefficient equal to that for a straight line, ignoring the point lying beyond the 'critical resolution'.

We do not say that graphs with linear segments (of different slopes) cannot exist. Indisputable examples have been produced, for example in rocky structures by Pape *et al.* (1985). Mandelbrot (personal communication, in Rigaut, 1984) would nonetheless have preferred, if our concavities should bear him out, to invoke successive values of D, rather than a slowly drifting dimension. Such a 'drift' (this is the word he used) is nevertheless invoked as a possibility by Mandelbrot (1982), for example for modelling the tributaries of rivers.

Having thus studied the literature, we found ourselves faced with an awkward dilemma: abandon the entire fractal idea or else find a new model. The concave graphs could not be explained by experimental error. Curvedness at weak resolutions can be explained by the use of certain algorithms for measuring the perimeter (such as the programme of Schwarz and Exner, 1980), but it is at strong resolutions that the concavity is most pronounced.

We decided that, after all, as a natural perimeter should not tend to infinity, and we did not much care for 'critical resolutions' and other curves broken into successive 'segments', the existence of an asymptote (maximal perimeter) to a globally concave log–log curve would indeed be quite normal. It remained to find a suitable function furnishing some theoretical justification.

14.4.2 A semi-fractal equation

Instead of a constant fractal dimension, it seems that, in most of the cases we have studied, there is a 'continuous drift of the fractal dimension', to which the concavities of the Richardson–Mandelbrot graphs bear witness. The corresponding objects will be called semi-fractals.

We have discovered empirically a really simple equation which can be used to study the variation of the perimeters (B_r) of objects as functions of the length (r) of the scale-step of the polygonation (which, as we have seen, expresses the inverse of the resolution). Our semi-fractal model, which will be based on this equation, will embody various things which will help to explain why it gives such an extraordinarily close fit with experimental observations. This first 'validation' of our equation resembles that of Richardson's formula (1961) by Mandelbrot (1967): in both cases, the justification for a formula discovered empirically is found in a fractal (or semi-fractal) model originally based on it.

Having noticed that a graph of B_r^{-1} as a function of r^c always yielded a straight line for a suitable value of the constant c (always with correlation coefficients > 0.95, for a wide variety of objects), we empirically discovered an equation that fitted the experimental data very well, as well as possessing interesting properties:

$$B_r = (1 + L^c r^c)^{-1} B_m \tag{1}$$

where B_m, c and L are constants. B_m is the (asymptotic) maximal value of B_r as $r \to 0$. L is the value of r^{-1} such that $B_r = B_m/2$.

There are two linearized versions of equation (1) that may be used to produce graphical or, better, iterative estimates of the parameters. The first is

$$B_r^{-1} = L^c B_m^{-1} r^c + B_m^{-1} \tag{2}$$

which exhibits analogies with the formula of the Lineweaver–Burk graph (1934) currently being used for enzymatic kinetics. The line, which we estimate by maximization of the correlation coefficient of linear regression, with $B_r^{-1} = f(r^c)$ (iteration on the value of c), croses the y axis at B_m^{-1} and the x axis at $-L^{-c}$. An initial rough estimate of c, before any iteration, is given by

$$\log [(B_{r_0} - B_r) B_r^{-1}] = \log (\mathrm{e}^{c \log r} - r_0^c) - \log (L^{-c} + r_0^c)$$

where r_0 is the smallest experimental value of r. The graph of $\log ([B_{r_0} - B_r) B_r^{-1}]$ against $\log r$ gives an increasing concave curve of gradient a:

$$a = c[1 - (r_0^{c} r^{-c})]^{-1} \tag{4}$$

This gradient generally changes slowly within a certain range of values of r and its estimate in this zone makes possible a first estimate of c in (4) by iteration.

The other linearized version of (1) is

$$\log [(B_m - B_r) B_r^{-1}] = c \log r + c \log L \tag{5}$$

One carries out iteration on B_m to maximize the correlation coefficient of linear regression and it is then an easy matter to determine c and L. This type of formula has been described by Wyman (1963), under the name 'Hill graph', for the study of the kinetics of allosteric enzymes. It seemed more robust (less sensitive in linear regression to straying values) to us than the first one. The parameters can be estimated with the help of a more sophisticated algorithm, used in enzymology (Atkins, 1973).

14.4.3 Interpretation of the equation—the semi-fractal model

Equation (1) can be written in the form:

$$B_r = [B_m^{-1} + (B_1^{-1} - B_m^{-1})r^c]^{-1} \quad (6)$$

where B_1 is the value of B_r for $r = 1$. When $B_m \rightarrow \infty$, $L \rightarrow \infty$ as well, and

$$B_r \rightarrow B_1 r^{-c} \quad (7)$$

This expression tends to resemble $B_r = B_1 r^{1-D}$, the Richardson–Mandelbrot equation. Our equation thus allows one to describe a semi-fractal outline as well as one of the 'classical' fractal variety. Note that, in the latter case, the Hill graph is unfeasible (one can nevertheless halt the iterations on B_m when it gets too big) and that of Lineweaver–Burk shows the line passing through (0, 0). The constant c is always less than 1 experimentally, as the theory suggests ($0 \leqslant c \leqslant 1$), since $c + 1 = D$ in the classical fractal case and we are dealing with outlines ($1 \leqslant D \leqslant 2$).

In the general case, as $B_r = N_r r$,

$$(B_m - B_r)L^{-c} = N_r r^{c+1} = Q_r \quad (8)$$

Thus, $c + 1$ is the dimension in which Q_r is the length of a fractal curve, if $0 < c < 1$. However Q_r drifts continually as r varies. When $B_m \rightarrow \infty$,

$$Q_r \rightarrow B_1 = N_r r^{c+1} \quad (9)$$

similar to $B_1 = N_r r^D$, where B_1 is a constant measure of B_r in a space of dimension D.

An interesting version of (1) is

$$\log B_r = -\log[1 + L^c e^{c \log r}] + \log B_m \quad (10)$$

For $B_m < \infty$, this certainly leads to a concave curve for the Richardson–Mandelbrot log–log graph (log B_r against log r. Its gradient is

$$b = -c(1 + L^{-c} r^{-c})^{-1} \quad (11)$$

and the curve is asymptotic to $\log B_m$ as $r \rightarrow 0$, and to

$$\log B_r = -c \log r + \log(B_m L^{-c}) \quad (12)$$

as $r \rightarrow \infty$. Note the analogy of (12) with the Richardson–Mandelbrot formula, so that, when $r \rightarrow \infty$, our equation tends to the latter. One can say that *the outline of a semi-fractal object tends to be ideally fractal at the weakest resolutions.* The measure of B_r in dimension $D(= c + 1)$ then tends to $B_m L^{-c}$.

Note that the limit, as $B_m \rightarrow \infty$, of

$$\Lambda_r = N_r r^{1-b} \quad (13)$$

where b is defined as in (11), is $N_r r^D$. Λ_r is a measre of the object in the dimension $(1 - b)$, but depends on r in the semi-fractal case. When the object is ideally fractal ($B_m \rightarrow \infty$ and $L \rightarrow \infty$), then $(1 - b) \rightarrow (c + 1) = D$, and $\Lambda_r \rightarrow B_1$.

Figure 2 shows the theoretical log–log graphs calculated from our equation,

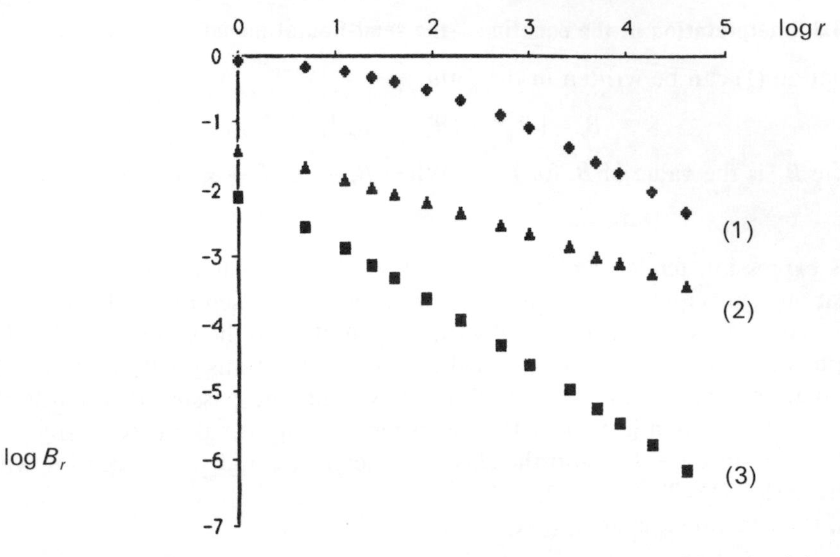

Figure 14.2 Theoretical log–log curves calculated from equation (1) of our semi-fractal model, varying the constants B_m, L and c.

(1) $B_m = 1$, $L = 0.1$, $c = 0.5$.
(2) $B_m = 1$, $L = 10$, $c = 0.5$.
(3) $B_m = 0.3$, $L = 1.5$, $c = 1$ ($\log B_m = -1.2$).

With the range of r employed, it is difficult not to take (2) and perhaps (3) for 'ideal' fractal lines and it is impossible to guess that the asymptote (B_m) of (3) of different from that common to (1) and (2). See the more detailed comments in the text

for different values of B_m, L and c. The left-hand part (weak values of r) is deliberately omitted. The reader should ask how such curves could be interpreted, using the classical fractal model, if they were obtained experimentally: (1) would present problems (impossible to fit a straight line); (2) would produce great joy—what a lovely straight line!—and still the horizontal asymptote is not far above (similar to (1)); (3), we are quite sure (as we have seen, one has only to read the literature), would often see itself being fitted to a straight line, at least over quite a large range of values of r, and this when its horizontal asymptote is close at hand (lower than in the other two cases)! Here are three semi-fractals of which at least one and maybe two would run the risk of being taken for pure fractals. Note that, for this graph, the scale of r goes from 1 to 1000: this, in pixels, is what we use in experimental practice, and corresponds to a range of resolutions that is quite common in biology.

All things considered, we believe that there is a continuous passage from the semi-fractal object to the pure fractal object, which represents its *ideal limit*. Moreover, at weak resolutions, those at which the topological character of the object (the 'finite product') can be observed, there is an asymptotic tendency to fractality. The outline of a semi-fractal object does not possess a constant fractal dimension: its dimension 'drifts' as a function of the resolution. Its perimeter does not tend to infinity at high resolutions, but to a maximal value; the more this value is (relatively) increased, the more the object can be acceptably regarded as approximately an ideal fractal.

14.5 The lung as a semi-fractal object

(With Bengt Robertson of the *Karolinska Institute, Stockholm*)

14.5.1 From the lung to steel via the geranium

Our equation fits remarkably well with experimental data of all kinds. Several examples of this are given in Rigaut (1984). The same values are estimated for the three parameters (B_m, L and c) when working, as well as from scale-steps, at different microscopic or macroscopic magnifications. The examples here studied range from the leaves of different varieties of geranium to cuts of steel and cast iron, and thence to histological sections of many different human and animal organs and cells. The correlation coefficients of linear regressions used in the estimation of the parameters of our equation are *always* > 0.95 and most often > 0.99. This clearly shows the interest of our model, which can be used even when non-linear log–log graphs rule out any serious estimation of a fractal dimension. The parameters of our equation turn out to be very useful for classifying different states of the same object (leaves of plants as a function of their age, or normal or pathological biological tissues, for example). We have particularly concentrated on lungs, and it is the results of our experiments in this area that we shall now present. The pulmonary structure is still far from being clearly described. We believe that our model throws some new light on the matter.

14.5.2 Alveolar outlines are semi-fractal

Figure 14.1 provided us with an example of a perimeter (B_r) of a cross-section of an object (studied by looking at microscopic sections) at different step-lengths of polygonation (r), tending, not to infinity, but to a limiting value (B_m) at high resolutions (small values of r). It related to the alveolar outlines of the lungs of prematurely born rabbits, underdeveloped (untreated) or well developed (treated with surfactant at birth). It was impossible, on the log–log graph, to estimate a fractal dimension (D) throughout the range of resolutions studied. It is nevertheless possible to estimate D at weak resolutions, because the graph is then acceptably linear; in these conditions, we find values of around 1.10 (well-developed lungs) and 1.25 (underdeveloped); the first are a little below those reported in the normal adult man (1.17) by Weibel (1979), but the pulmonary structure does continue to get more complex after birth.

Our equation fits the experimental data very closely (all the correlation

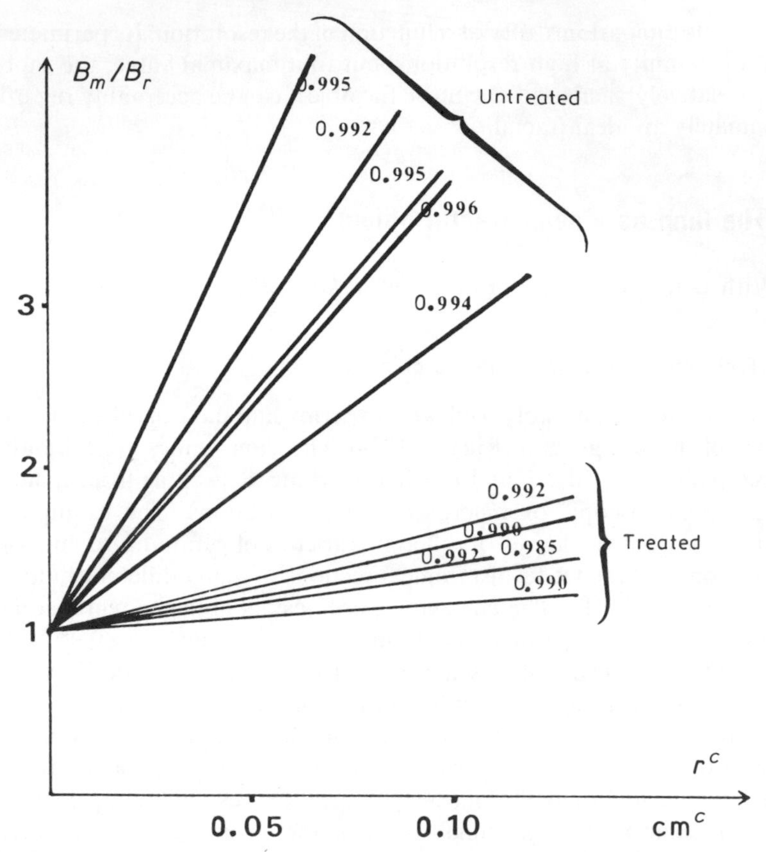

Figure 14.3 Results obtained on lungs; see Table 14.1 for the experimental conditions (only the 60% groups are represented here.) The Lineweaver–Burk graph following our equation (1). Excellent separation (compare with Figure 14.1) between the untreated (underinflated alveoles) and animals treated with surfactant (well-expanded alveoles). The lines of linear regression have been estimated with averages of B_r^{-1} on the five microscopic fields studied. Note the very high correlation coefficients (the figures shown). The estimates of the parameters by this method (iteration on c) are very close to those (Table 14.1) obtained by Hill's method (iteration on B_m)

coefficients of linear regression are >0.99). Figure 14.3 shows the regression lines for both the untreated and treated cases, obtained from the Lineweaver–Burk-type expression of our equation; the spread-types between fields, not shown for the sake of clarity, were very weak. Our model gives a better separation between treated and untreated (compare Figures 14.1 and 14.3).

Table 14.1 shows the values of the semi-fractal parameters estimated with the help of our model. The constant L can be used for a very clear classification of the four experimental groups (treated and untreated, each with 20 or 60% of inspiration time with respect to the respiratory cycle; Rigaut *et al.*, 1985a); the Wilcoxon–Mann–Whitney U-test is an indicator of statistical significance of the differences between pairs of these groups, except for the pair untreated

Table 14.1 Experimental results obtained on microscopic sections of the lungs of prematurely born rabbits, badly developed (untreated) or well developed (treated with surfactant); 20 and 60% (of inspiration times with respect to the respiratory cycle) correspond to two distinct experimental conditions (Rigaut and Robertson, 1985). From top to bottom of the table, the groups correspond to animals having, on average, better and better expanded pulmonary alveoles. An automatic image analyser (IBAS) was used, with an algorithm for the polygonation of alveolar outlines by steps of variable size. Five microscopic fields were studied for each section. Estimation, by Hill's method (see text), of the parameters of our equation (1) (averages and spread types between animals; $R=$ correlation coefficient, by linear regression). The constant L allows good separation of the experimental groups.

Groups		*n*	*R*	$S_{mV}(\mathrm{cm}^{-1})$	*e*	$L(cm^{-1})$
Untreated	20%	5	0.994	747	0.63	685
			(0.003)	(46)	(0.20)	(471)
	60%	5	0.994	672	0.61	357
			(0.002)	(102)	(0.07)	(210)
Treated	20%	5	0.993	595	0.46	126
			(0.002)	(98)	(0.09)	(33)
	60%	5	0.990	503	0.34	54
			(0.002)	(56)	(0.08)	(33)

20%/60% (and this only seems to be due to the presence of a section of very heterogeneous structure amongst the 20%'s). The only significant difference for the constant c is found for the pair untreated/treated 60%. $S_{mV}(=(4/\pi)B_{mA})$ differs significantly for the pairs untreated/treated 20 and 60%, but S_m does not in fact show any difference (this is the reference volume V which clearly increases when the lung is inflated). In short, there is clearly a progressive gradation of the results from the untreated up to the treated 60% (optimal treatment). This is absolutely parallel to that observed for stereological parameters other than $S_V:V_V$, $CV(V_V)$ and K_V(see later) (Rigaut *et al.*, 1983a; Rigaut and Robertson, 1985). The high values of the spread types of Lin the untreated cases reflect the heterogeneity of their lungs ($CV(V_V)$ is also very high in such cases; Rigaut and Robertson, 1985). The estimates were made using the Hill-type linearized version of our equation, by iterations on B_{mA}. The densities of length of the alveolar outlines per unit reference area (B_{rA}) were evaluated by the image analyser with scale-steps of from 1 to 100 pixels (with 2.3 pixels/μm).

14.5.3 A semi-fractal object possesses outlines with infinitely self-similar convolutions

If it is true that the alveolar surface area tends to a maximum at high resolution, the number of points of inflexion (I) of its cross-sectional outlines, on the other hand, tends to infinity. This fact, which may seem strange at first sight, is in fact perfectly reasonable. To see this, it suffices to imagine an ever-increasing number of smaller and smaller waves; their number will tend to infinity, but, under certain conditions, their total length can tend to a maximal limit. We shall see that our semi-fractal 'monsters' furnish typical examples of this.

Figure 14.4 shows a log–log graph of the number of points of inflexion per unit length of alveolar outline (I_r/B_r) against the step-length (r). This graph appears as a broken line; the right-hand part is, probably artificially, of different slope, due to an edge effect (inflexions are overlooked, particularly at large

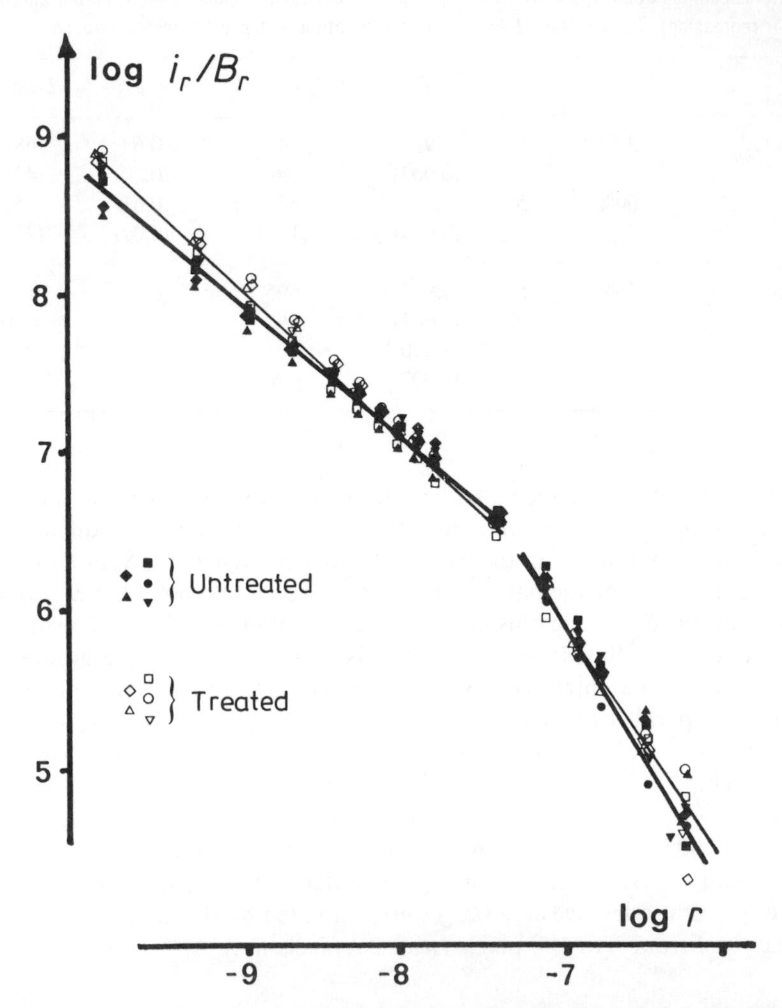

Figure 14.4 Results obtained on lungs; see Table 14.1 for the experimental conditions (only the 60% groups are represented here). The number of points of inflexion per unit length of alveolar outline (I_r/B_r) is independent of the degree of pulmonary development, for each 'resolution' (step-length r). The quantity I_r/B_r reflects the density of mean integral curvature of the asymptotic lines of the saddle surfaces per total area of alveolar surface (see text). The saddle surfaces represent elements of stability ('anchorages') of the alveoles at each resolution. Moreover, the number of points of inflexion seems to tend to infinity at stronger resolutions

step-lengths, because the edges of the field of the microscope intersect certain alveoles). The left-hand part has a slope of -1. This figure is almost precisely what is found for all the objects we have studied so far! An interpretation in terms of our semi-fractal monsters will be offered in due course.

14.5.4 A new model for the alveolar structure

It is quite striking (Figure 14.4) that all the lungs studied, well developed or underdeveloped, have the same I_r/B_r for each r. On the other hand, when I_r/B_r is replaced by I_{rA}, the untreated then have a greater density of points of inflexion per reference area than the treated. The explanation of these findings is simple: the alveolar inflation does not 'rub out' the inflexions, but spreads them apart (in stretching the alveolar outlines), and so the density per unit length of outline remains constant while that per unit reference area diminishes. We have measured the angles of rotation of the tangents to the convex and concave arcs of the outlines between the points of inflexion: they diminish just when the alveoles are well inflated. Indeed, as $I_A = \frac{1}{2}K_{sV}$, where K_{sV} is the volume density of the mean integral curvature of the asymptotic lines of the saddle surfaces (DeHoff, 1981), we have with $S_V = (4/\pi)B_A$:

$$K_{sS} = \frac{\pi}{2} I_B \tag{14}$$

where K_{sS} is the density of the mean integral curvature of the asymptotic lines of the saddle surfaces per unit area of total alveolar surface (Rigaut *et al.*, 1983b, 1985a). It is thus this new stereological quantity, K_{sS} (or K_{srS}, for a given step-length r), which remains constant, independently of the degree of alveolar development (since S is constant, the lung is not very elastic). Our interpretation in structural terms is that the alveolar system admits, at each resolution, a framework of fixed regions, 'anchorages', really, represented by the saddle surfaces, the asymptotic lines of which would form rigid supporting networks. As this holds good at each resolution, the model we propose is, from this point of view, infinitely self-similar.

14.5.5 A semi-fractal object can exhibit a discontinuity in its structure

Figure 14.5 shows the experimental log–log graph of the average mean integral curvature, or the mean integral curvature averaged over the surface area ($\bar{K} = K_S = K_V/S_V$; Cahn, 1967; DeHoff, 1967), as a function of the step-length ($\bar{K}_r$ as a function of r). The curvature seems to tend to infinity at stronger resolutions. A fractal outline has no curvature, strictly speaking; it is, however, legitimate to consider the estimated curvature values for discrete resolutions. Note that the mean alveolar curvature is thus almost constant (particularly on well-developed lungs) for r between 2.7 and 12.3 μm. Maybe this fact will one day be linked with physiological notions (gaseous exchanges)? K is estimated by the formula $K_V = \pi T_{\text{net}A}$, where T_{net} is the difference between the number

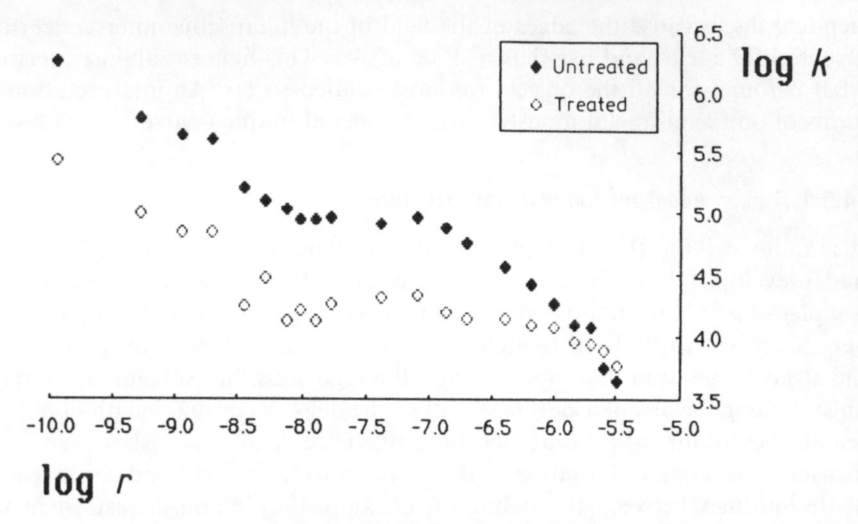

Figure 14.5 Results obtained on lungs; see Table 14.1 for the experimental conditions (only the 60% groups are represented here). The quantity $\overline{K}$, the average mean integral curvature (the mean integral curvature averaged over the area of the alveolar surface), as in an 'ideal' fractal, seems to tend to infinity at strong resolutions, but shows a plateau in a certain range of resolutions

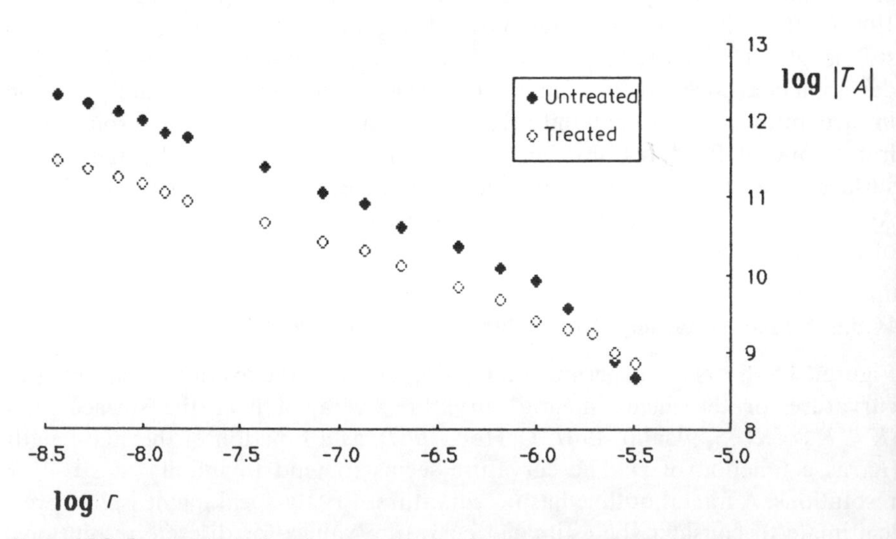

Figure 14.6 Results obtained on lungs; see Table 14.1 for the experimental conditions (only the 60% groups are represented here). The quantity $|T|$ is the absolute number of horizontal tangents to the alveolar outlines. For Figure 14.5, K was evaluated from T_{net}, the difference between the positive horizontal tangents (convexities) and negative (concavities) to the outlines. $|T|$ seems to tend to infinity, which is consistent with the same behaviour for I (Figure 14.4)

of positive (convex edges) and negative (concave edges) horizontal tangents to the alveolar outlines (Cahn, 1967; DeHoff, 1967). What the image analyser evaluates is thus T_{net} ($T_{\text{net}r}$, for each r, here). The other curve is that of $|T_A|$ (the total, rather than the difference, of the positive and negative tangents); this has no flat part (Figure 14.6).

14.5.6 Curvatures—relevant or not?

A semi-fractal object thus displays a finite surface area, but with strictly self-similar 'twists' (inflexions); this we have observed for a wide variety of objects. Nevertheless, certain spatial quantities, such as the integral curvature, can exhibit non-self-similar relationships in certain ranges of resolution. We repeat that, strictly speaking, there is no curvature for such objects, since they are infinitely convoluted; all the same, their curvature at a given resolution represents an interesting approximation (the 'curvature' is then that of a curve which fits with the outline at this resolution). For lungs the average mean integral curvature ($\bar{K}$) is related to physiologically interesting quantities, such as the alveolar surface tension, $\gamma(p = 2\gamma\bar{K}$, Gibbs' law, where p is the alveolar pressure; Weibel, 1979), and the mean alveolar radius, $R(R = 2\gamma/p$, Laplace's law, where $R = \bar{K}^{-1}$). Such relationships take on a completely new significance in the fractal setting: the alveolar radius and the surface tension should not vary with the resolution when $\bar{K}$ varies! If one chooses a given value of $\bar{K}$, at what resolution should one do it?

14.6 A gallery of semi-fractal monsters

14.6.1 Motivation

As Mandelbrot (1982) has splendidly shown, with undeniable artistry, the more or less simple representations of fractal monsters such as the von Koch island and other Peano curves help bring the concepts of fractal geometry to life. In this spirit, we should find new guidelines in an attempt to start up a new, and as yet empty, collection: that of semi-fractal monsters.

14.6.2 A curious coincidence?

The perimeter of the von Koch island (see Mandelbrot, 1982) tends to infinity as one goes on adding new and ever-smaller triangles indefinitely. What about its area (A), however? It certainly tends to a limit (A_m), but how? Writing down the formula for $\log(A_m - A_n)$, replacing n (the stage of the construction) by its value, and using logs, in $r_n = r_1(1/3)^{n-1}$ (where r_n is the side of a triangle at the nth stage), we get

$$A_a = [1 - \tfrac{3}{8}(a/a_1)^{1-D/2}]A_m \tag{15}$$

where A_a is the area of the island at the stage of the construction corresponding to the addition of triangles of area a (a_1 is the area of the triangle at the beginning). D is the fractal dimension of the outline of the island ($D = 2\log 2/\log 3$). When $a \ll a_1$, we have

$$A_a \approx (1 + L^c a^c)^{-1} A_m \qquad (16)$$

and one will observe the similarity between this formula and equation (1) which is the basis for our semi-fractal model. Here, $A_m = (8/5)a_1$, $c = 1 - D/2$ and $L = (3/8)^{1/c}$.

14.6.3 A rudimentary semi-fractal monster

Take a square of side r_0. Inscribe in one of its corners a square of side $r_1 = r_0/2$. Then add a new one, always keeping the same ratio of its side to that of its predecessor ($r_n = r_{n-1}/2$), with one of its corners opposite to the 'free' corner of the first one, and continue in this way. One easily finds that the total perimeter, at the stage where one has constructed a square of side r, is

$$B_r = 2r_0[1 - 2(r/r_0)] \qquad (17)$$

When $r \ll r_0$, we have

$$B_r \approx (1 + L^c r^c)^{-1} B_m$$

i.e. again the equation of our empirical law (1), with $B_m = 2r_0$, $c = 1$ and $L = 2/r_0$. Slightly changing the rule for the construction, one can obtain different values for B_m and L. All this, however, is a long way from a representative model of an outline of a natural object!

14.6.4 The semi-fractal monster... of Archimedes

This is obtained by polygonating, with a number (N_n) of equal straight line segments of length r_n, the inside of a circle of radius R, starting with a square of side r_1, then drawing an octagon, and carrying on in this way ($N_n = 2^{n+1}$). We are interested in the perimeter of the polygon at each stage of the construction (B_n):

$$B_n = 2R\, 2^{n+1} \sin(\pi/2^{n+1}) \qquad (18)$$

Of course, $B_n \to B_m = 2\pi R$, as $n \to \infty$. Following two restricted series expansions, we find that, when $r \ll r_1$:

$$B_r \approx (1 + L^c r^c)^{-1} B_m$$

i.e. we are back with our equation (1), with, here, $B_m = 2\pi R$, $c = 2$ and $L = (\pi/12R)^{1/c}$. However, this monster is not what we are really looking for, because we would like convoluted outlines and, above all, the possibility of transition to a pure fractal by modifying the parameters.

14.6.5 Towards 'coherent' semi-fractal monsters

Here are some ground rules for coherent constructions (the terminology has been explained in the preceding sections; C is a positive constant (<1)):

$$B_n = N_n r_n, \qquad \begin{array}{c} n \to \infty \\ B_n \to B_m \quad (B_n < B_m) \\ N_n \to \infty \end{array}$$

$$r_n = Cr_{n-1} = r_1 C^{n-1}, \qquad r_n \to 0$$

Each stage is thus simply deduced from its predecessor. Perhaps Nature develops in this way (the branching patterns of natural tree-like structures suggest it)? The ingredients of the outline (straight lines or curves) will have a self-similar length deducible from one stage to the next. This will serve to define this class of 'coherent' monsters. We will thus have

$$B_n = r_1 C^{-1} N_n C^n$$

and $N_n C^n$ should $\to \max(> Cr_1^{-1})$, as $n \to \infty$. N_n needs to be defined in such a way that this condition is satisfied.

A first type of 'coherent' semi-fractal monsters

This will be defined as follows:

$$N_n = N_1 \left(aM^{mn} - \sum_{p=0}^{m-1} w_p M^{pn} \right) \qquad (w_p \geqslant 0)$$

$$r_n = r_1 M^{-m(n-1)} \qquad (M, m, n = \text{positive integers})$$

It can be shown that, in this case, when n becomes very large,

$$B_r \approx (1 + L^c r^c)^{-1} B_m$$

i.e. once again our equation (1), with $B_m = N_1 r_1 M$, $c = m^{-1}$ and $L = r_1^{-1} M^{-1}$.

If the number of points of inflexion I_n (whatever the geometric representation chosen for the monster) is proportional to N_n, which would seem reasonable (N_n 'construction units', of waves for example), one shows that the slope of the log–log graph of I_r against r tends to -1 as $r \to 0$. Recall that a slope of this value has been observed experimentally for all the objects we have studied.

The Fibonacci semi-fractal monsters

These will be defined by

$$N_n = U_{2mn} - \sum_{p=0}^{m-1} w_p U_{2pn} \qquad (w_p \geqslant 0; m \geqslant 1)$$

where

$$U_{2n} = 5^{-1/2}(K^{-2n} - K^{2n}) \qquad (K = (5^{1/2} - 1)/2)$$

defines the numbers U_i for even numbers i of the Fibonacci sequence. With

$$r_n = r_1 K^{2m(n-1)}$$

we again find, when $r_n \ll r_1$, our equation (1):

$$B_r \approx (1 + L^c r^c)^{-1} B_m$$

with, when $w_{m-1} > 0$, $c = m^{-1}$, as for the previous type of monsters.

If the number of points of inflexion of the geometric curve of the outline of the monster is proportional to N_n, one shows that the slope of $\log I_r$ against $\log r$ here again tends to -1 as $r \to 0$.

14.6.6 What do our monsters resemble?

At present, they are only given by equations.... (Artistic?) inspiration has so far failed us in our search for a satisfactory geometric representation, i.e. one which, most importantly, admits coherent successive subdivision. All sorts of designs are possible, but they have so far failed to possess the good regularity of the 'classical' fractal curves.

One could well ask if the replacement of specific functions by probabilistic elements might not better correspond with our semi-fractal concept. True fractality represents, in our view, let us repeat, an ideal towards which a more chaotic scheme could then converge.

It nevertheless came as a pleasant surprise when we established that the above equations all tended to the equation of our empirical law.

14.7 The frustration of the little quadrat

—Variance of dispersion of the area density and fractals

(With Christian Lantuejoul of the *Centre de Géostatistique et de Morphology Mathématique, Fontainebleau*)

This section tells the story of a twofold frustration: our own, when we realized that the equations derived from geostatistical theory did not allow us to explain our experimental findings, and that of the little quadrat... which offered us the solution!

14.7.1 A curious experimental fact

We were interested in the variance of the area density ($\mathrm{var}(a_A)$) of objects distributed over a surface, as a function of the area(A) of a typical quadrat, a family of which, all equal, completely pave the surface being studied. This type of variance (with a complete paving of a surface S by quadrats s) is called 'variance of dispersion'. As the size of the quadrats decreases, the variance certainly tends to some constant value: the point variance ($\mathrm{var}(a_A)_m$).

We have found experimentally, with all kinds of objects, that the log–log graph of $\mathrm{var}(a_A)$ against A turns out, in almost all cases, to have a linear part

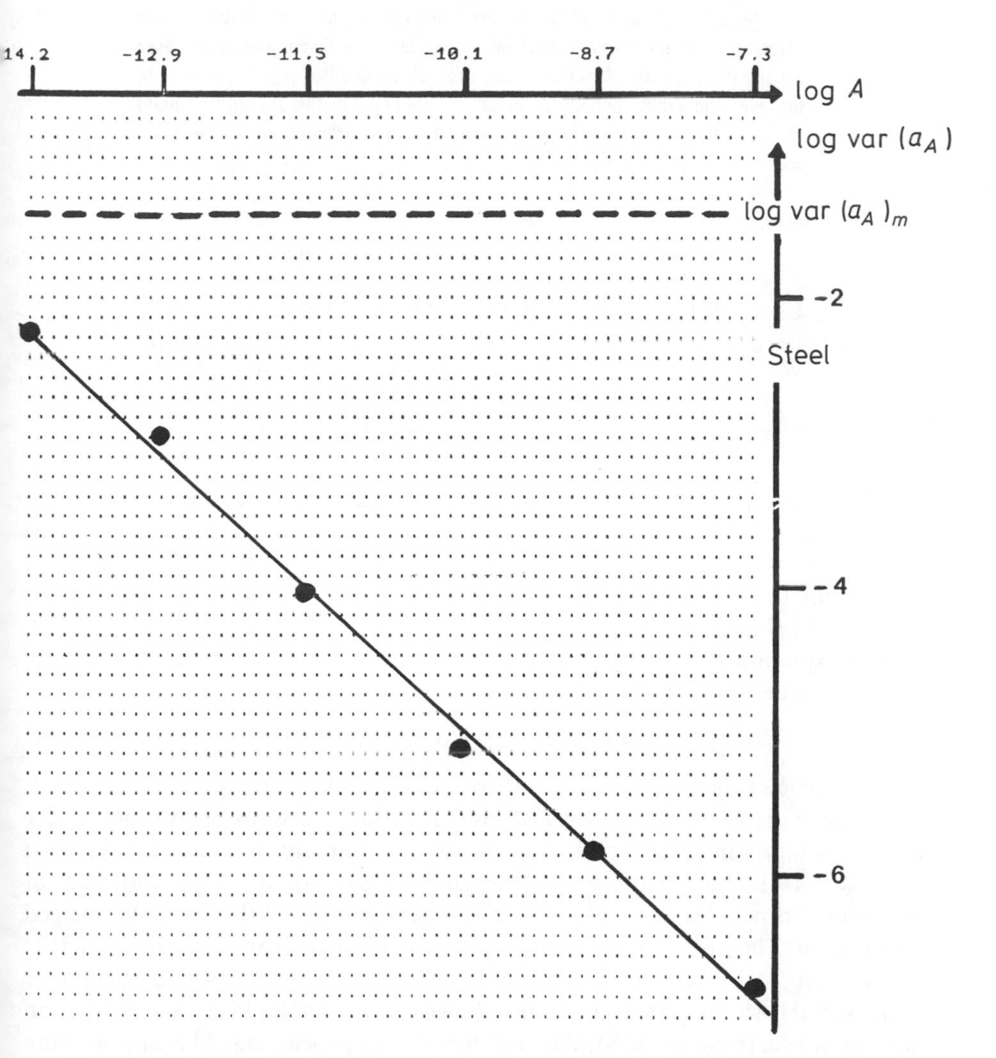

Figure 14.7 Experimental result for a micrograph of steel sheet. Very similar results have been obtained for all the objects studied, including the lung; we have chosen steel here to emphasize the generality of our models, which do not have to be restricted to biology. The variance of the area density (var (a_A)) between quadrats (squares) is studied with quadrats of different areas (A), always with a complete paving of 10 microscopic fields by quadrats ranging from 16^2 to 512^2 pixels. Automatic image analysis (IBAS). Var (a_A) evidently tends to the point variance (var $(a_A)_m$) for small quadrats, but a decidely linear part of the graph is to be seen here, as in the majority of cases. As var (a_A) reflects the perimeter of the outlines of objects (B_r) at different resolutions (see text), this fact is not at all surprising (the same linear portion is most frequently seen for log B_r against log r (see Figure 14.1))

Table 14.2 Results obtained on lungs; see Table 14.1 for the experimental conditions (only the '60%' groups are represented here). Estimation of the parameters of our equation (19) for the variance of dispersion of the area density ($\mathrm{var}(a_A)$) as a function of the area of the quadrat (A). The areas are those of the cross-sections of alveoles. $\mathrm{Var}\,(a_A)_m$ is the theoretical point variance. Correlation by linear regression (Hill's method), using the point variance as upper asymptote.

Groups	*n*	$\mathrm{var}(a_A)_m$	*b*	*M* (cm^{-2})	*R*
Underinflated	5	0.230	0.48	2376×10^3	0.997
		(0.039)	(0.05)	(1815×10^3)	(0.002)
Well inflated	5	0.212	0.46	505×10^3	0.999
		(0.040)	(0.04)	(225×10^3)	(0.001)

for the largest quadrats; very often (Figure 14.7), this linear portion extends to the smallest quadrats considered (of course, below this, the curve becomes concave as it tends to $\mathrm{var}(a_A)_m$). The form of these graphs irresistably calls to mind that which has become familiar for the perimeters of objects as a function of the step-length (Figure 14.1).

Our equation (1) fits the experimental points (Table 14.2) remarkably well, when rewritten in the form:

$$\mathrm{var}\,(a_A) = (1 + M^b A^b)^{-1}\,\mathrm{var}\,(a_A)_m \tag{19}$$

The objects studied were very varied: polished slices of steel sheet, cast iron plates, hepatic knots on sections of rat's liver, lung sections from prematurely born rabbits (untreated and treated with surfactant, see above), confetti (round—'14th July' variety or triangular—cut haphazardly) scattered at random and pebbles from a beach on the west of Jutland (Denmark) arranged at random. The area measurements, in the quadrats of variable size (16^2–512^2 pixels), on 10 fields were carried out with the help of an automatic image analyser (IBAS) together with an algorithm in FORTRAN-80 calling on subroutines written in ASSEMBLER for matrix processing. The objects were digitized after being photographed by a television camera, and then binarized according to degrees of greyness.

14.7.2 Theoretical considerations in the setting of the stationary model

To begin with, we tried to interpret out experimental results in the setting of the stationary model of order two. The basic tool in this case, much used in geostatistics, is the variogramme:

$$\gamma(h) = \tfrac{1}{2}\,\mathrm{var}\,(Z_{x+h} - Z_x) = C(0) - C(h)$$

When h is large, the correlation between the outcomes at x and at $x + h$ tend

Table 14.3 Results obtained on lungs; see Table 14.1 for the experimental conditions (only the '60%' groups are represented here). Estimation of the parameters of the variogramme of type $\gamma(h) = K|h|^{\beta}$ (correlation by log–log linear regression). K is estimated on the basis of h being in centimeters.

Groups	*n*	β	*K*	*R*
Underinflated	5	0.52	5.5	1.000
		(0.08)	(3.8)	(<0.001)
Well inflated	5	0.77	26.0	1.000
		(0.07)	(10.7)	(<0.001)

to 0 and $\gamma(h)$ approaches an asymptote when $h \rightarrow \infty$. The value (h_0) at which this asymptote is regarded as being sufficiently close is the 'range'. The variance of dispersion can be expressed as a function of the variogramme:

$$\operatorname{var}(s|S) = [A(S)]^{-2} \int_S \int_S \gamma(x-y)\,\mathrm{d}x\,\mathrm{d}y - [A(s)]^{-2} \int_s \int_s \gamma(x-y)\,\mathrm{d}x\,\mathrm{d}y$$

Three cases should be considered:

(a) According to the order of the size of both s and S.
We can hypothesize (Table 14.3) that

$$\gamma(h) = K|h|^{\beta}$$

The variance of dispersion then takes the form

$$\operatorname{var}(s|S) = [1 - A(s)A(S)^{-1}]^{\beta/2} C(0)$$

where $C(0)$ is the point variance. If s is small,

$$\operatorname{var}(s|S) \approx [1 + (A(s)A(S)^{-1})^{\beta/2}]^{-1} C(0) \quad (20)$$

a formula of the same type as our semi-fractal equation for outlines (1) and as that established experimentally for $\operatorname{var}(a_A)$ (19).

(b) According to the order of the size of s alone.
In this case, one shows (Rigaut *et al.*, 1985b) that, when s is small,

$$\operatorname{var}(s|S) \approx [1 + K\mathscr{I}_{\beta} C(0)^{-1} (A(s))^{\beta/2}]^{-1} C(0) \quad (21)$$

which is again of the same form as (1) and (19), with

$$\mathscr{I}_{\beta} = \int_0^1 \int_0^1 (a^2 + b^2)^{\beta/2} (1-a)(1-b)\,\mathrm{d}a\,\mathrm{d}b$$

(c) When the sizes of s and S are both much largest larger than the range.
One then shows (Rigaut *et al.*, 1985b) that

$$\operatorname{var}(s|S) \approx [(A(s))^{-1} - (A(S))^{-1}] \int_{\mathbf{R}^2} C(h)\,\mathrm{d}h \quad (22)$$

14.7.3 The stationary model is not applicable to our observations

Our experimental conditions correspond to case (b) above. One should then find experimentally, with our equation (19), an M^b near to $K.\mathscr{I}_\beta C(0)^{-1}$, obtained from equation (21) and the value found analytically for $\mathscr{I}_\beta$ under our experimental conditions. We have systematically determined K and β in using the variogrammes.

In fact, we found a difference of several orders of magnitude between the theoretical constant and that determined experimentally! Moreover, the constant b is not always equal to $\beta/2$ as the theory would like.

To explain such differences, one could question the limits of validity of the probabilistic model. Even if it is true that the nice homogeneous image of the steel sheet can be regarded as the realization of a stationary aleatory set, this is probably not the case for lungs, for example. Furthermore, one must proceed with care when using formulae based on approximations for the variance of dispersion. The behaviour of the variogramme at the origin is, moreover, very hard to estimate well experimentally. Lastly, the size of the quadrats must lie within a range specified by the limits of validity of the approximation formulae; this last point could explain why differences exist even for the steel sheet. Nevertheless, they are so huge for all the objects studied that we are forced to question the validity of the model employed. We have therefore come up with a totally different model.

14.7.4 An explanation of fractal type: the 'frustrated little quadrat'

Why fractal geometry?

We were drawn to the use of the concepts of fractal geometry by a variety of facts: (a) equation (19), derived from (1), fits the experimental data (Table 14.2) very closely; (b) our variogrammes have a type $K|h|^\beta$ behaviour at the origin, with $\beta < 1$ (Table 14.3, for lungs), and this implies that the perimeter of the object is infinite and therefore that the object is fractal; (c) a relationship between β and the fractal dimension (D), in the context of the Weierstrass–Mandelbrot model ($\beta = 4 - 2D$), has been conjectured (Berry and Lewis, 1980); Burrough, 1981) and confirmed in the case of brownian motion (Mandelbrot and Wallis, 1969; Mandelbrot, 1982; Burrough, 1984).

The model, intuitively speaking

Imagine an object with self-similar outlines. To best visualize what follows, imagine that the outlines, at every resolution, look like waves, of fairly regular wavelength and sufficiently high compared to this. Imagine the (binary) object in a field (S) that is paved with small quadrats (s), in order to determine the variance of dispersion with the image analyser. In each quadrat, the area density (a_A) is measured. Var(a_A) will be particularly high (i.e. will tend to the point variance) when a great many quadrats only comprise pixels of the object ($a_A = 1$) or of the surrounding background ($a_A = 0$). Var(a_A) will be particularly low when very many quadrats comprise pixels of both the object and the background.

All things considered, it is thus the probability that a quadrat overlaps the outline of the object that is the decisive factor. This will occur for every quadrat whose centre lies within the zone formed by the enlargement of the outline by a square structuring element (Serra, 1972, 1982) of the same size as the quadrat. This zone (Figure 14.8), a Minkowski sausage, is defined in a similar way to that produced by enlargement of the outline by means of circles, which, as we have seen, can serve to estimate the perimeter for each 'resolution', the inverse

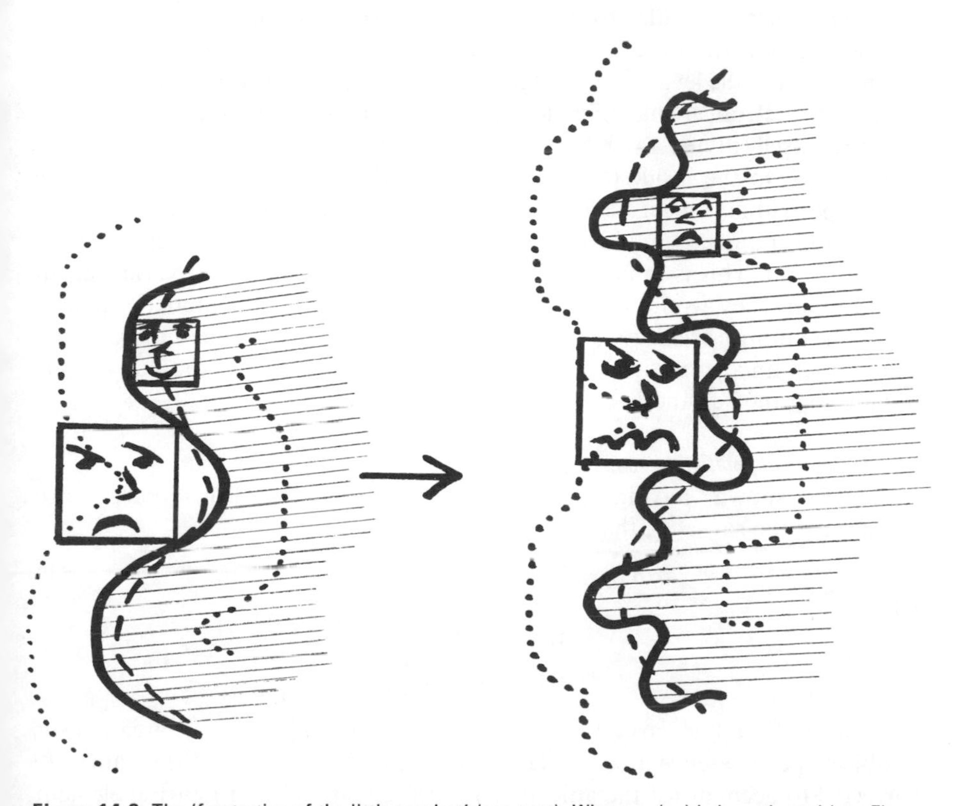

Figure 14.8 The 'frustration of the little quadrat' (see text). When var (a_A) is investigated (see Figure 14.7) as a function of A (the area of the quadrat), the determining factors are the density of the outlines and their convolutedness at each resolution, for the more quadrats straddle the outlines of objects, the weaker is the variance. The corresponding probability is a function of the relative area of the Minkowski sausage obtained by enlargement of the outlines using the quadrat as the structuring element (indicated by the dotted lines for the larger of the two quadrats shown). This sausage obtained by enlargement is analogous to that used for determining the fractal dimension of outlines the broken line curves represent outlines 'modelled' at the resolution which allows exploration of the sausage). Referring to the left-hand diagram, the larger quadrat cannot 'insinuate' itself into the waves of the outlines, but the smaller one can do it better. On the right-hand side, the same outline is observed at a more powerful magnification and at the thus increased resolution new waves appear (fractal or semi-fractal object). Our quadrats are 'frustrated' at being unable, while having the same absolute size, to fit themselves any more closely to the digitations of the albeit enlarged outlines. Our method of the quadrats and of the variance of the area density leads, if the object is fractal, to an easy estimate of the fractal dimension of its outlines

of which is given by the radius of the circle. The square is plainly non-isotropic, but this can be taken into account by means of a correction.

It is thus clear that the variance of dispersion is a function of the perimeter (B_r), estimated, at the 'resolution' corresponding to the size of the quadrat, by the sausage method. The convolutedness of the outline, which determines the thickness of the sausage for a given size of quadrat, therefore plays a major role. This seems capable of explaining why one recovers, in this context, an empirical equation similar to the one that we proposed for B_r.

In addition, when a quadrat is not able to 'fit' itself entirely within (or outside of) the object at the level of some digitation of the outline, it will be... 'frustrated' to find, at a stronger magnification, new digitations in which it will never be able to fit itself (Figure 14.8)! Moreover, a smaller quadrat, which adapted itself well enough to the inside (or to the outside) of comparatively 'roomy' twists and turns, will no longer be able to do it at a higher degree of magnification.

These phenomena should help to reduce the variance of dispersion for a given quadrat. This is in fact what we observe: the experimental variances are weaker than they would be in the setting of the stationary model (equation (21)). The outlines are those of a semi-fractal object and, for each quadrat, there exist twists and turns into which it cannot get and is 'frustrated' at thus diminishing the variance!

The model, mathematically speaking

It can be shown, having regard to the above observations (particularly concerning the sausage), that

$$\mathrm{var}(a_A)_r = N(N-1)^{-1}(W_1 - W_2 r' B_{r'A}) \tag{23}$$

where

$$W_1 = \overline{a_A}(1 - \overline{a_A})$$
$$W_2 = (\overline{q_Q}) - (\overline{q_Q^2})$$

and where $\mathrm{var}(a_A)_r$ is $\mathrm{var}(a_A)$ for a quadrat of side r, $\bar{a}_A$ the mean area density of objects per reference area, N the total number of quadrats, r' the value of r corrected to account for the anisotropy of the square as a structuring element, $B_{r'A}$ the outline length density estimated on the basis of the area of the sausage ($\Delta = r'B_{r'}$; this estimate of $B_{r'}$ is equivalent to that obtained by means of a polygonation with a step-length r'), $(\overline{q_Q})$ the mean area density in the quadrats which include at least one pixel of the object and $(\overline{q_Q^2})$ the mean square area density in the same quadrats. We rediscover (a classical result) that the point variance is

$$\mathrm{var}(a_A)_m = N(N-1)^{-1} W_1 \tag{24}$$

It is easy to show that $(\overline{q_Q}) = \frac{1}{2}$ when the 'half'-portion of the sausage situated inside the object has area $(\Delta/2)$. A closed object which is not very big relative to the quadrat will yield a value of $(\overline{q_Q}) < \frac{1}{2}$. It can be shown that $(\overline{q_Q^2}) = (27/60) - [(7\log 2 - 1)/15\pi]$ and that $(\overline{q_Q}) = \frac{1}{2}$, when the intersection of the outline with the surface of the quadrat is always a straight line on integrating over all its positions and angles. The value of W_2 will depend on the mean integral curvature of the object at the 'resolution' of the quadrat (i.e. of the outline after smoothing out by closure with the quadrat as the structuring element).

Putting $r' = Cr$, A (area of the quadrat) $= r^2$ and

$$K = CW_1^{-1}W_2$$

we have

$$\mathrm{var}(a_A)_r = \mathrm{var}(a_A)_m(1 - KB_{r'A}A^{1/2}) \tag{25}$$

Note that, for a circular outline (and maybe this is true, at least under certain conditions, for every isotropic outline), $C = 4/\pi$.

Using our equation (1) to express $B_{r'}$ and considering the case of a small quadrat (more precisely, when $KB_{r'A}A^{1/2}$ is $\ll 1$), we get

$$\mathrm{var}(a_A)_r \approx (1 + KB_{mA}A^{-1/2})^{-1}\,\mathrm{var}(a_A)_m \tag{26}$$

which is very similar to that found empirically (19), with $b = \frac{1}{2}$ and $M = K^2B_{mA}^2$.

At the present time, our experimental estimates compare favourably with those that may be deduced from (26) on taking, for K and C, the values seen above.

A new method for determining the fractal dimension

When the classical fractal equation ($B_{r'} \approx \lambda r'^{1-D}$) is applicable, then, for the little quadrats, (19) always holds, but with

$$b = 1 - D/2$$
$$M = (C^{2-D}W_1^{-1}W_2\lambda)^{1/b}$$

and it is easy to determine the fractal dimension of the outlines once the constant b has been evaluated experimentally by the method of quadrats and the variance of dispersion of the area density of objects:

$$D = 2(1 - b)$$

Our method thus enables D to be determined without ever being directly involved with the outlines!

Curvature and resolution

A fractal object has no curvature, strictly speaking. Nevertheless, as we have seen, it can be defined for each resolution. Its study by means of the coefficient W_2, as indicated above, can thus be contemplated.

A final interesting consequence

It is possible, with our quadrat method, to obtain an estimate for the length of outline(s) of an object(s) at different 'resolutions', starting from $\mathrm{var}(a_A)$

evaluated with quadrats of different sizes: in fact, one obtains, from (25),

$$B_{r'A} = K^{-1}(1 - [\mathrm{var}(a_A)_r \, \mathrm{var}(a_A)_m^{-1}])A^{-1/2} \qquad (27)$$

Since this equation is valid whatever the model (fractal or semi-fractal) of variation of $B_{r'}$, it is therefore possible to determine experimentally which model to choose, for example using our equation (1).

Some comments

We believe this line of research has a great future. By relating the variance of dispersion, the variogramme and a (brownian?) model of the outline, it should be possible to create a new tool for characterizing the fractality of an edge of an object.

Furthermore, it should be possible to generalize our model to multigrey (and not just binary) images. Instead of outlines predefined in an unambiguous way, one should then get involved with 'outlines' defined in terms of shades of grey. Here, the outlines will not be predefined by the researcher, and this seems more natural to us than defining them with the help of a range of greys in order to binarize something regarded *a priori* as an 'object'. The object would then be the outcome and no longer the starting-point of an investigation. One may dream: might it not be possible, in this way, iteratively weighing the contributions of the quadrats to the variance, to arrive at a model for the *recognition* (rather than the analysis) of forms of given fractality?

14.8 Discussion

The concept of fractal structures often involves the notion of internal similarity, with a constant fractal dimension that embodies a measure of the object studied which is independent of the resolution. Everybody agrees with the idea that, in the real world, the length of an object's outline cannot tend to infinity and therefore that a limit ('cut-off') must be reached as the resolution increases (i.e. with polygonation by step-length, when the latter tends to zero). This limit has been interpreted as being a 'critical resolution', beyond which the structure is seen in all its complexity (Paumgartner *et al.*, 1981; Mandelbrot, 1982). We believe that there is no clear-cut limit, but rather a progressivity—a drift of the fractal dimension.

The existence of successive fractal dimensions has been proposed, accompanied in the Richardson–Mandelbrot log–log graphs by segments of broken lines; this would explain the existence of several successive structures, even 'structural' and 'textural' fractals, constituting 'multifractals' (Kaye, 1977). The existence of concave log–log graphs is well established; certain authors had observed this, and our experiments with a wide variety of objects confirm it beyond doubt. We have not seen graphs with broken lines (or with several concavities), but some fine examples where several linear and concave parts occur alternately have been reported for rocky structures by Pape *et al.* (1985).

Moreover, these authors have successfully used our semi-fractal model to parameterize the concave parts; the range of resolutions employed went from the atomic to the visible. One should clearly distinguish the cases where several distinct zones, linear and/or concave, exist from those where it is not necessary to 'invent' broken lines that do not exist; a statistical study on this subject would be really useful.

When there is a concavity, we introduce a continuous transition of the fractal dimension: from ≈ 1 at high resolution, it tends progressively, in our model, to $D(=c+1)$ at weak resolution. A linear segment is often to be found towards the right-hand side of the log–log graph, where the concave curve, according to our theory, tends to an oblique asymptote of slope $1-D(=-c)$. Our equation tends to that of Richardson–Mandelbrot at weak resolution.

Certain experimental objects yield a truly linear graph; looking at the range of resolutions employed, it seems that they are generally a long way off (i.e. much weaker than) those that would reveal the detailed structure: the coast of Brittany yields a linear graph, but it is observed from an aircraft—so who is going to measure it using the microscope?

We believe that Nature employs semi-fractal constructions (structural, observable at high resolution), but that these tend to a 'final result' (topological, at weak resolution) of fractal type: fractality is the geometric ideal approached by such structures.

The equation that underlies our fractal/semi-fractal model was obtained empirically. This is also true of that of Richardson (1961), who only appreciated its true value following its fractal interpretation by Mandelbrot (1967). In linguistics, Mandelbrot (1953, 1968) mathematically derived the 'Zipf–Mandelbrot law' for word-frequencies, which expresses itself by means of a log–log concavity (probability $p(n)$ of occurrence of words in a text, as a function of their position n in a list arranged in order of decreasing frequency): $p(n)=(m+n)^{-\beta}F$, where F, m and β are constants. This law fits the experimental data better than Zipf's empirical one (1949): $p(n)=n^{-1}$, which could only accommodate the observations linearly on a log–log graph. Mandelbrot (personal communication, in Rigaut, 1984) pointed out to us that we could have used, instead of our own, another interpolation equation, which displays analogies with that of Zipf–Mandelbrot: $B_r=(1+Lr)^{-c}B_m$; on the one hand, its properties are a little different from those of our own ($B_r=(1+L^c r^c)^{-1}B_m$) and, on the other hand, β^{-1} (and not β) can be interpreted, in the context of the Zipf–Mandelbrot law, as being a fractal dimension (Mandelbrot, 1977, 1982), while in our context the variable dimension $c+1$ tends to a constant dimension D at weak resolution.

Our observations have been carried out with an automatic image analyser—a computer—and therefore with digitized images. The validity of our interpretations has nevertheless been verified manually on certain photographs of objects and, moreover, the same estimated values for the parameters of our equation are obtained by the method of variable step-lengths, but still using several degrees of magnification. The study of Pape *et al.* (1985) in petrology has demonstrated the validity of our model throughout very large ranges of resolutions, when log–log concavities exist. These authors have succeeded in correlating the parameters of our model with the petrophysical data. Lastly, they have shown that our equation also fits in very accurately with the variations of areas as a function of the resolution; apropos of this, remember that the area of the von Koch island can be modelled by our equation.

We think that our study of the variances of dispersion of the area-densities as a function of the areas of quadrats holds much promise for the future. We have described a new method for determining the fractal dimension of an outline. A generalization of our concept of quadrats to multigrey images is on the way to being achieved.

The use of fractal models in biology is still too rare. This will certainly change in the years to come, but it would be advisable to clearly separate the (rare?) cases where there is a constant fractal dimension, at least in a given range of resolutions, from those where one should work in a semi-fractal context. Our model allows a good parametrization in all cases, purely fractal or otherwise.

The scientific study of shape is still in an incoherent infancy. If it is true that structure influences shape, fractal geometry has a great future in biology and in materials science. Painters have always known that the structure (the substance) ought to be at the root of the 'shape' (the final result). Crudely drawn lines only result in bad paintings and the young child, as yet untrammelled by our rigid topological (ideological) straitjacket, understands the soft outlines and the creative textures of things. When, therefore, will people stop concerning themselves just with the boundaries of so-called 'exact' objects, rather than with how these vary with resolution, which would enable them to understand the generating structures of observed shapes?

References

Atkins G. L. (1973) A simple digital-computer program for estimating the parameters of the Hill equation, *Eur. J. Biochem.*, **33**, 175–180.

Barenblatt G. I. and Monin A. S. (1983) Similarity principles for the biology of pelagic animals, *Proc. Nat. Acad. Sci., USA.*, **80** 3540–3542.

Berry M. V. and Lewis Z. V. (1980) On the Weierstrass–Mandelbrot fractal function, *Proc. Roy. Soc. A,* **370**, 459–484.

Bignon J. and André-Bougaran J. (1969) Mesure automatique de la surface alvéolaire du poumon de rat à l'aide d'un microscope téléviseur quantitatif, *C.R. Acad. Sci., Paris,* D/**269**, 409–412.

Bolender R. P., Paumgartnér D., Losa G., Mullener D. and Weibel E. R. (1978) Integrated stereological and biochemical studies on hepatocyte membranes—1. Membrane recoveries in subcellular fractions, *J. Cell Biol.*, **77**, 565–583.

Burrough P. A. (1981) Fractal dimensions of landscapes and other environmental data, *Nature*, **294**, 240–242.

Burrough P. A. (1984) The application of fractal ideas to geophysical phenomena, *Bull. Inst. Math. Appl.*, **20**, 36–42.

Cahn J. W. (1967) The significance of average mean curvature and its determination by quantitative metallography, *Trans. Metall. Soc. AIME*, **239**, 610–616.

Chermant J. L. and Coster M. (1978) Fractal object in image analysis, in *Proc. Int. Symp. Quant. Metallography*, Florence, Ass. ital. di Metallurgia, pp. 125–137.

Chermant J. L., Coster M. and Lavole J. (1981) A study of the ductile–brittle transition in steels using quantitative fractography, in *Proc. 3rd Eur. Symp. on Stereology*, Ljubljana. Stereol. Iugoslav., Vol. 3, Suppl. 1, pp. 225–232.

Coster M. and Deschanvres A. (1978) Fracture, objet fractal et morphologie mathématique, in J. L. Chermant (Ed.), *Quantitative Analysis of Microstructures in Materials Science, Biology and Medicine*, Prakt. Metall., Special Issue 8, pp. 61–73.

DeHoff R. T. (1967) The quantitative estimation of mean surface curvature, *Trans. Metall. Soc. AIME*, **239**, 617–621.

DeHoff R. T. (1981) Stereological meaning of the inflection point count. Stereology 5, *J. Microsc.*, **121**, 13–19.

Flook A. G. (1978) The use of dilation logic on the Quantimet to achieve fractal dimension characterisation of textured and structured profiles, *Powder Technol.*, **21**, 295–298.

Flook A. G. (1979) The characterization of textured and structured particle profiles by the automated measurement of their fractal dimensions, in *Proc. PARTEC 2nd Eur. Symp. on Particle Characterization*, Nuremberg, pp. 591–599.

Flook A. G. (1982) Fractal dimensions: their evaluation and their significance in stereological measurements, in *Proc. 3rd Eur. symp. on Stereology*, 2nd part, *Acta Stereol.*, **1**, 79–87.

Flook A. G. (1984) A comparison of quantitative methods of shape characterization, *Acta Stereol.*, **3**, 159–164.

Fredberg J. J. and Hoenig A. (1978) Mechanical response of the lungs at high frequencies, *Trans. ASME, J. Biol. Engng.*, **100**, 57–66.

Gehr P., Bachofen M. and Weibel E. R. (1978) The normal human lung. Ultrastructure and morphometric estimation of diffusion capacity, *Respir. Physiol.*, **32**, 112–140.

Jean R. V. (1983) Allometric relations in plant growth, *J. Math. Biol.*, **18**, 189–200.

Kaye B. H. (1977) Characterization of the surface area of a fine particle profile by its fractal dimension, in M. J. Groves (Ed.), *Proc. Salford Conf. on Particle Size Analysis*, Heyden, pp. 250–259.

Kaye B. H. (1978) Specification of the ruggedness and/or texture of a fine particle profile by its fractal dimension, *Powder Technol.*, **21**, 1–16.

Keller H. J., Friedli H. P., Gehr P., Bachofen M. and Weibel E. R. (1976) The effect of optical resolution on the estimation of stereological parameters, in *Proc. 4th Int. Cong. Stereology*, Gaithersburg, US Dept. of commerce, Nat. Bureau of Standards, Special Publ. **421**, pp. 409–410.

Lefèvre J. (1983) Teleonomical optimization of a fractal model of the pulmonary arterial bed, *J. Theor. Biol.*, **102**, 225–248.

Lefèvre J. and Barreto J. (1983) Reduced and teleonomical models in cardiovascular physiology, in G. C. Vansteenkiste and P. C. Young (Eds.), *Modelling and Data Analysis in Biotechnology and Medical Engineering*, North Holland, pp. 27–37.

Lefèvre J., Barreto J. and Gonze P. (1982) Teleonomical representation of arterial beds by fractal symmetric and asymmetric trees, in A. Ballester, D. Cardus and E. Trillas (Eds.), *Proc. 2nd World Conf. on Maths at the Service of Man*, Las Palmas, pp. 435–443.

Lefèvre J., Barreto J., Gonze P., Dawant B. and Alexandre X. (1983) Optimal features of arterial systems: input impedance, in *Proc. 36th ACEMB*, Columbus, 40.4, p. 205.

Lineweaver H. and Burk D. (1934) The determination of enzyme dissociation constants, *J. Am. Chem. Soc.*, **56**, 658–666.

Loehle C. (1983) The fractal dimension and ecology, *Specul. Sci. Technol.*, **6**, 131–142.

Loud A. V. (1968) A quantitative stereological description of ultrastructure of normal rat liver parenchymal cells, *J. Cell Biol.*, **37**, 27–46.

Lovejoy S. (1982) Area–perimeter relation for rain and cloud areas, *Science*, **216**, 185–187.

Mandelbrot B. B. (1953) Contribution à la théorie mathématique des communications, Thèse Doct. ès Sci. Math., Publ. Inst. Stat. Univ. Paris, Ser. A, Vol. 2, No. 2521.

Mandelbrot B. B. (1967) How long is the coast of Britain? Statistical self-similarity and fractional dimension, *Science*, **156**, 636–638.

Mandelbrot B. B. (1968) Les constantes chiffreés du discours, in J. Martinet (Ed.), *Linguistique. Encyclopedie de la Pleiade*, Gallimard, Paris, pp. 46–56.

Mandelbrot B. B. (1977) *Fractals. Form, Chance and Dimension*, Freeman, San Francisco.

Mandelbrot B. B. (1982) *The Fractal Geometry of Nature*, Freeman, San Francisco.

Mandelbrot B. B. and Wallis J. R. (1969) Computer experiments with fractional gaussian noises, *Water Res.*, **5**, 228–267.

Mandelbrot B. B., Passoja D. E. and Paullay A. J. (1984) Fractal character of fracture surfaces of metals, *Nature*, **308**, 721–722.

Matheron G. (1965) *Les Variables Régionalisées et leur Estimation*, Masson, Paris.

Morse D. R., Lawton J. H., Dodson M. M. and Williamson M. H. (1985) Fractal dimension of vegetation and the distribution of arthropod body lengths, *Nature*, **314**, 731–733.

Olah A. H. (1980) Effects of microscopic resolution on histomorphometrical estimates of structural and remodelling parameters in cancellous bone, *Path. Res. Pract.*, **166**, 313–322.

Orford J. D. and Whalley W. B. (1983) The use of fractal dimension to quantify the morphology of irregular shaped particles, *Sedimentology*, **30**, 655–668.

Pape H., Riepe L. and Schopper J. R. (1985) Theory of self-similar network structures in rock materials and their investigation with microscopical and physical methods, in *Proc. 4th Eur. Symp. on Stereology*, Gothenburg, *Acta Stereol.* (in press).

Paumgartner D., Losa G. and Weibel E. R. (1981) Resolution effect on the stereological estimation of surface and volume and its interpretation in terms of fractal dimensions. Stereology 5, *J. Microsc.*, **121**, 51–63.

Reith A. (1973) The influence of triiodothyronine and riboflavin deficiency on the rat liver with special reference to mitochondria. A morphologic, morphometric and cytochemical study by electron microscopy, *Lab. Invest.*, **29**, 216–223.

Reith A. (1976) Is there an unrecognised systematic error in the estimation of surface density of biomembranes? in *Proc. 4th Int. Cong. on Stereology*, Gaithersburg, US Dept. of Commerce, Nat. Bureau of Standards, Special Publ. **431**, p. 427.

Reith A., Barnard T. and Rohr H. P. (1976) Stereology of cellular reaction patterns, *Crit. Rev. Toxicol.*, **4**, 219–269.

Richardson L. F. (1961) The problem of contiguity: an appendix to statistics of deadly quarrels, *Gen. Systems Yearbook*, **6**, 139–187.

Rigaut J. P. (1984) An empirical formulation relating boundary lengths to resolution in specimens showing 'non-ideally fractal' dimensions, *J. Microsc.*, **133**, 41–54.

Rigaut J. P. and Robertson B. (1985) Quantitative evaluation of neonatal lung expansion with automated image analysis, *Ped. Pathol.* (in press).

Rigaut J. P., Berggren P. and Robertson B. (1983a) Automated techniques for the study of lung alveolar stereological parameters with the IBAS image analyser on optical microscopy seections, *J. Microsc.*, **130**, 53–61.

Rigaut J. P., Berggren P. and Robertson B. (1983b) Resolution-dependence of stereological estimations: interpretation, with a new fractal concept, of automated image analyser-obtained results on lung sections, in *Proc. 6th Int. Cong. on Stereology*, Gainesville, *Acta Stereol.*, **2** (Suppl. 1), 121–124.

Rigaut J. P., Berggren P. and Robertson B. (1985a) Stereology, fractals and semi-fractals—the lung alveolar structure studied through a new model, in *Proc. 4th Eur. Symp. on Stereology*, gothenburg, *Acta Stereol.* (in press).

Rigaut J. P., Lantuéjoul C. and Deverly F. (1985b) Relationship between variance of area density and quadrat area—interpretation by fractal and random models, in *Proc. 4th Eur. Symp. on stereology*, Gothenburg, *Acta Stereol.* (in press).

Schwarz H. and Exner H. E. (1980) The implementation of the concept of fractal dimension on a semi-automatic image analyser, *Power Technol.*, **27**, 207–213.

Serra J. (1972) Stereology and structuring elements, *J. Microsc.*, **95**, 93–103.

Serra J. (1982) *Image Analysis and Mathematical Morphology*, Academic Press, London.

Singhal S., Henderson R., Horsfield K., Harding K. and Cumming G. (1973) Morphology of the pulmonary arterial tree, *Circ. Res.*, **33**, 190–200.

Thurlbeck W. M. (1967) The internal surface area of non-emphysematous lungs, *Am. Rev. Resp. Dis.*, **95**, 765–772.

Underwood E. E. (1961) Discussion on measurement of interlamellar spacing of pearlite, *Trans. AIME*, **54**, 743–750.

Weibel E. R. (1963) *Morphometry of the Human Lung*, Springer-Verlag, Berlin.

Weibel E. R. (1979) *Stereological Methods*, Vol. 1: *Practical Methods for Biological Morphometry*, Academic Press, London.

Weibel E. R., Stäbuli W., Gnägi H. R. and Hess F. A. (1969) Correlated morphometric and biochemical studies on the liver cell. 1. Morphometric model, stereologic methods and normal morphometric data for rat liver, *J. Cell Biol.*, **42**, 68–91.

Wright K. and Karlsson B. (1983) Fractal analysis and stereological evaluation of microstructures, *J. Microsc.*, **129**, 185–200.

Wyman J. (1963) Allosteric effects in hemoglobin, *Cold Spring Harbor Symp. Quant. Biol.*, **28**, 483–489.

Zeeman E. C. (1977) *Catastrophe Theory. Selected Papers, 1972–1977*, Addison-Wesley, Reading, Mass.

Zipf G. K. (1949) *Human Behavior and the Principle of Least Effort*, Addison-Wesley, Cambridge, Mass.

Chapter 15 Reconstruction of images from projections

N. de Beaucoudray, L. Garnero and J.-P. Hugonin

Institut d'Optique, Université Paris-Sud

We are concerned with the problem of image formation in a spectral domain where lenses are not available (X-, γ-, α-rays). We are therefore dealing with a 'non-conventional' imagery.

From the physical point of view, these phenomena are more straightforward than classical optics since:

(a) There is no diffraction of the waves.
(b) There is not even any refraction.

The rays propagate in straight lines and are only attenuated by matter. (Figure 15.1).

This simplified physical situation is counterbalanced by the complicated mathematics involved in calculating the images.

Important remark Even if optical systems were available for the wavelengths that concern us, they would not meet our needs, because the objects concerned are not of the same kind as those encountered in photography. We are talking about objects in three dimensions (such as, for example, particles in suspension in a liquid) that can only be properly described by means of tomographic sections. A source emits radiation. This radiation crosses an obstacle before being detected by a receiver (Figure 15.2).

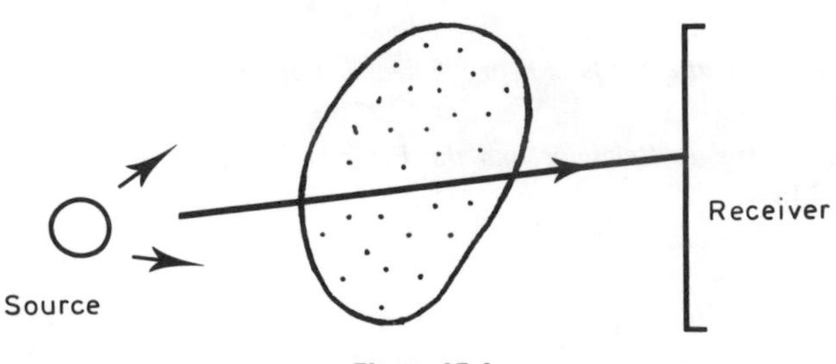

Figure 15.1

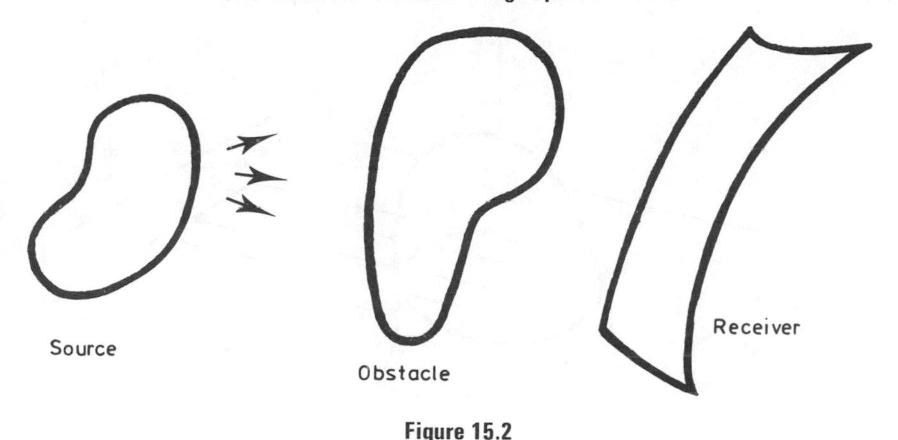

Figure 15.2

Two cases can be identified according to whether the object of study is the source or the obstacle.

(a) *Emission.* The obstacle is known and one seeks to determine the source. This, for example, is the case with imagery by code-breaking in nuclear medicine or with pin-hole imagery.
(b) *Absorption.* The source is known and one seeks to determine the obstacle; an example is the X-scanner.

15.1 Data-collection methods using separate sections

In transverse axial tomography, the object rotates about an axis, and the method of recording information does not mix up information coming from the different planes of sections perpendicular to this axis.

15.1.1 Absorption

The object is illuminated by a beam of parallel rays and the receiver registers the *shadow cast* by the object lit by this beam (Figure 15.3). If the object is characterized at each point by a coefficient of absorption μ, a ray which has traversed the object is attenuated by a factor of $e^{-\int\mu}$, where the integral is taken along the ray, which is, of course, a straight line.

Taking the logarithm of the received intensity, one measures $\int\mu$.

15.1.2 Emission

The emitter object (for example an organ with a radioactive trace attached to it) is characterized at each point by an intensity of radiation I (Figure 15.4). A collimator chooses a direction within the radiation so that, at each point of the

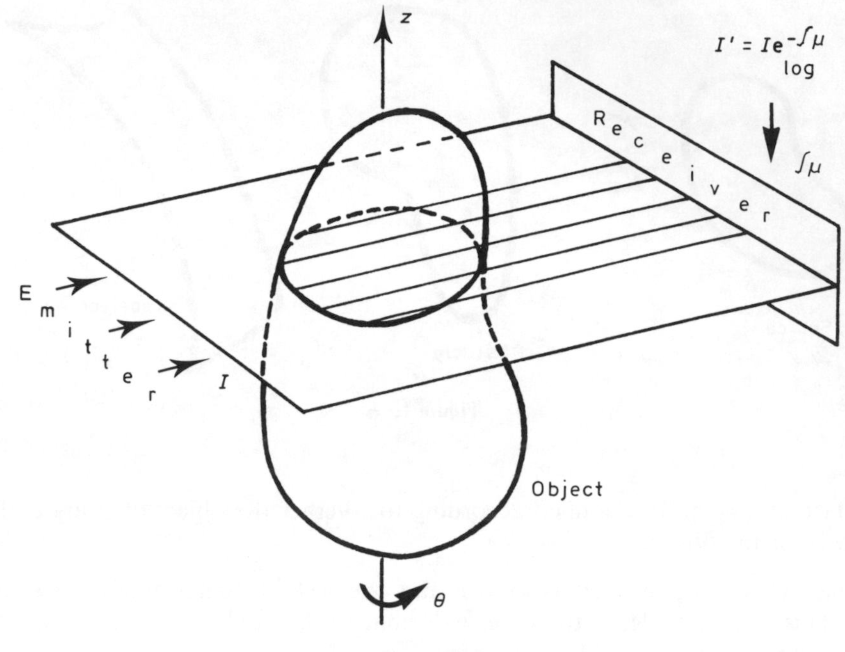

Figure 15.3

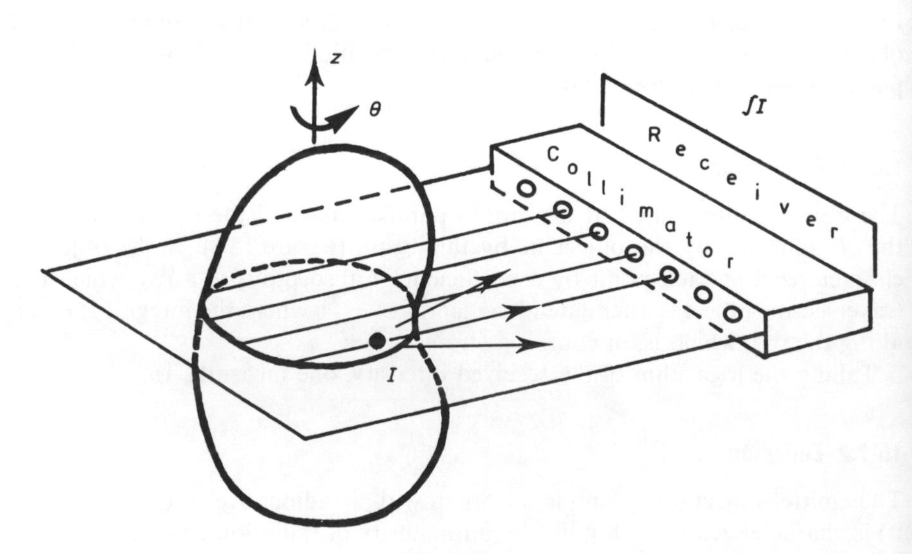

Figure 15.4

receiver, $\int I$ is recorded, where the integral is taken along the straight line passing through this point parallel to the direction of the collimator.

In the two cases, one varies the direction of the illuminating beam or the direction of the collimator by rotating the object about an axis. This is also equivalent to causing either the emitter–receiver assembly or the collimator–receiver assembly to rotate about this axis. One finds that sections perpendicular to the z axis register on different lines of the receiver. Reconstruction of the object in three dimensions thus comes down to the reconstruction of independent two-dimensional sections.

15.1.3 Projections

In the two cases, one records, at each point of the receiver, the integral along the line parallel to the axis of emission or of collimation of a function characterizting the object (the coefficient of absorption to the radiation or the density of the emitter body as appropriate).

The measurements thus provide, for each angle θ, the projections (Figure 15.5):

$$P_\theta(x') = \int f(x', y')\,\mathrm{d}y' \qquad \text{in the } (x', y') \text{ plane}$$

inclined at an angle θ to the fixed (x, y) plane.

15.1.4 The principle of the reconstruction

The one-dimensional Fourier transform of P_θ is equal to the two-dimensional Fourier transform of f in the θ direction of the Fourier plane (see Figure 15.6).

Being given the projections for all angles thus allows $\tilde{f}$ to be determined

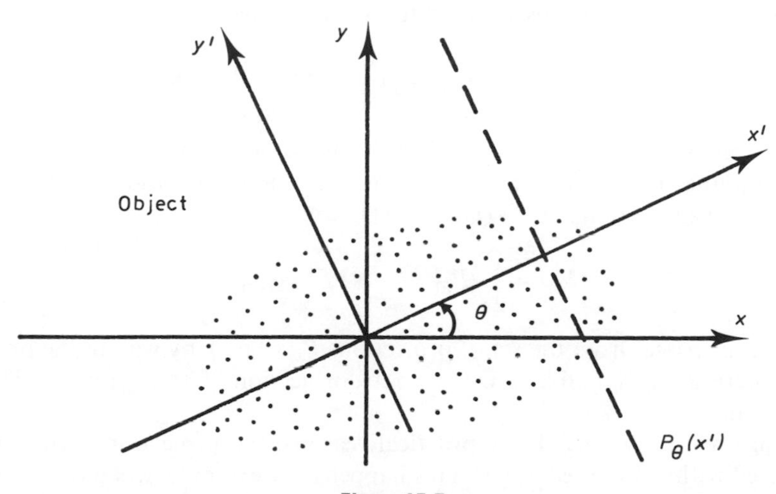

Figure 15.5

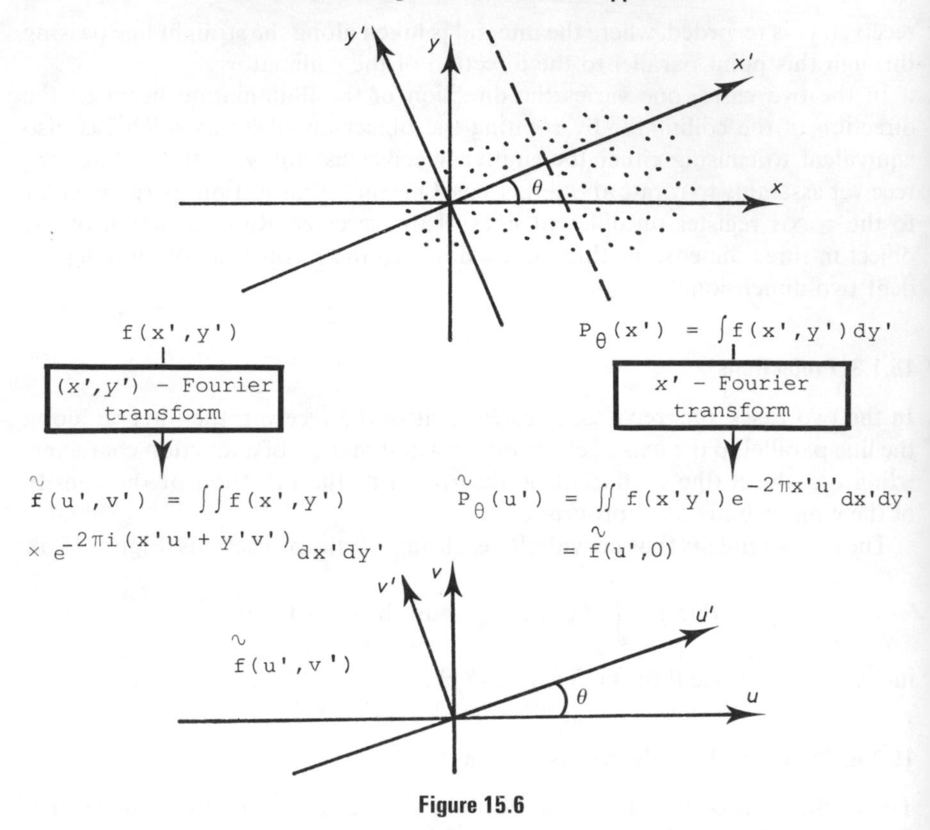

Figure 15.6

throughout the plane and hence f can be recovered by means of a two-dimensional inverse Fourier transform.

In polar coordinates, this Fourier transform is given by

$$f(M) = \int_0^{\pi} \left[\int_{-\infty}^{\infty} \tilde{P}_\theta(u')\, \mathrm{e}^{2\pi i u' x'} |u'|\, \mathrm{d}u' \right] \mathrm{d}\theta$$

The u' integral, with θ fixed, is in fact the one-dimensional Fourier transform of the product of $P(u')$ and u', and is therefore the convolution of P with the Fourier transform of u', i.e. $-(1/2\pi^2)\, Pf\,(1/x'^2)$:

$$f(M) = \int_0^{\pi} P_\theta \left(-\frac{1}{2\pi^2}\, Pf \left(\frac{1}{X'^2} \right) \right) \mathrm{d}\theta$$

One thus passes from the set of projections $P_\theta(x')$ to f by way of the product of a filtering (with θ fixed) with a retroprojection (filtered retroprojection algorithm).

Things will go wrong if, for practical reasons, the projections can only be measured within a limited angle. This happens, for example, when one is trying to get an image of the heart: the thoracic cage restricts the viewing angle.

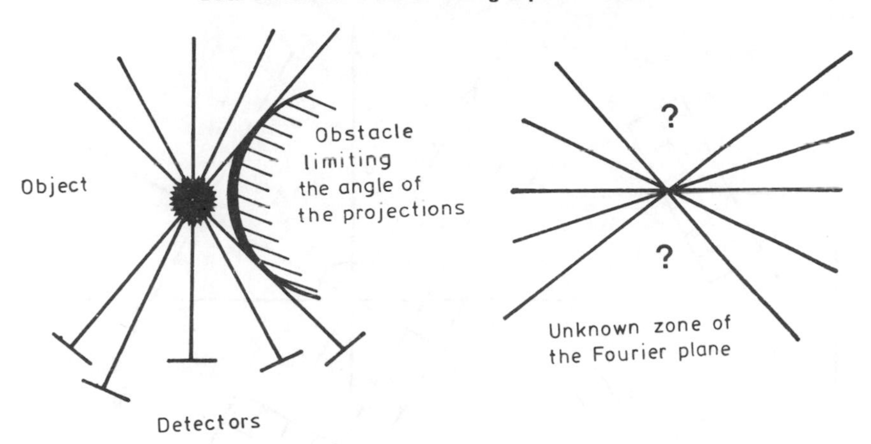

Figure 15.7

The preceding calculations show that part of the Fourier plane stays out of reach.

The filtered retroprojection algorithm does not lend itself easily to the introduction of *a priori* data. This has led us to tackle the problem of the reconstruction of the function f in a totally different way.

The *a priori* data will be expressed in the form of a mask $W(x, y)$ describing an approximate form of the object. For example, W could be the indicator function of the support of the object. The formalism is so constructed that the image obtained is better, the closer W is to the actual object. (Figure 15.7).

The physical data are always the projections. There are M directions of projection, with N' measurement points for each direction of projection, and thus a total of $N = M \times N'$ measured values. These values are rearranged into an N-dimensional vector P. More precisely, the projections may be expressed in the form:

$$P_i = \iint f(x, y)\, h_i(x, y)\, \mathrm{d}x\, \mathrm{d}y$$

where h_i is the indicator function of the band seen by the detector element (Figures 15.8 and 15.9).

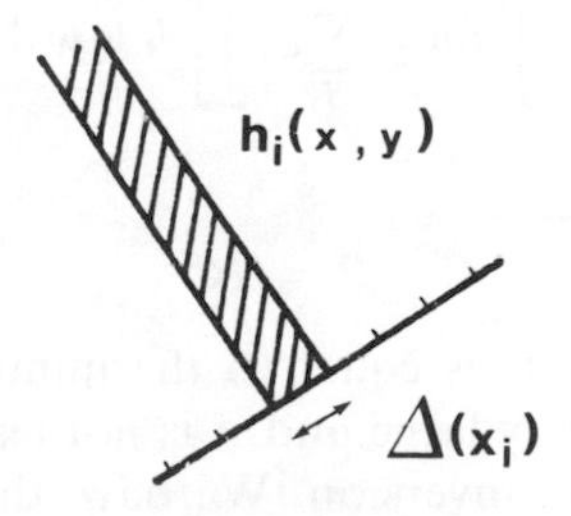

Figure 15.8

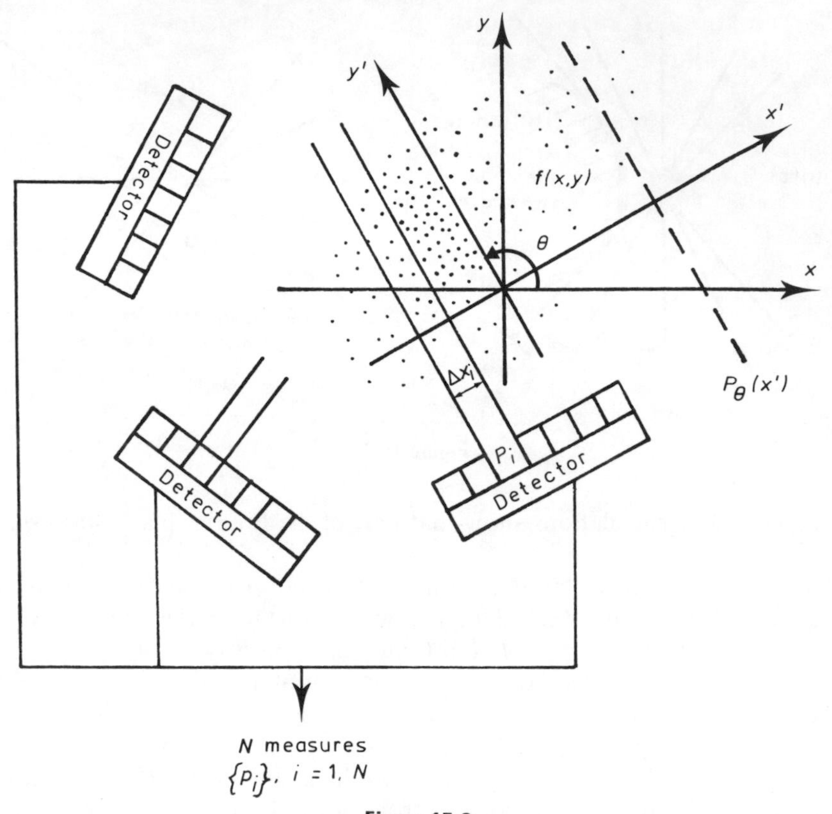

Figure 15.9

Define φ_i as the product of h_i with W. The originality of our method lies in seeking an approximation $\hat{f}$ to f in the form of a linear combination of the φ_i:

$$\hat{f} = \sum_i c_i \varphi_i$$

The image elements φ_i are thus directly linked to the data-collection geometry.

It suffices to express that the projections of $\hat{f}$ are the measured P_i to obtain a system of N equations connecting the N unknowns c_i:

$$P_i = \int \hat{f} h_i = \sum_j c_j \iint h_j h_i w \, dx \, dy$$

and in matrix form:

$$P = HC$$

The order of the matrix H is equal to the number of points in the image. This number is generally very large and it is not feasible to solve the equation $P = HC$ by a simple matrix inversion. We have therefore preferred to use an iterative method.

15.2 Tomographic reconstruction of images using the three-dimensional Radon transform

15.2.1 Emission

The flow from the radioactive object goes through a collimator before reaching the detector. The collimator consists of parallel planes and the detector is now linear (Figure 15.10).

Each detector element sees the object through a single plate of the collimator and thus collects the radioactive activity on this plane.

One thus obtains a series of projections of the object on the vector $\mathbf{\Omega}$ perpendicular to the planes of the collimator.

15.2.2 Another example

Another example of the use of the three-dimensional Radon transform occurs in certain types of RMN imagery. Without going into these phenomena in detail, suffice it to say that, if nuclei characterized by a gyromagnetic ratio γ placed in a magnetic field B are submitted to a radiofrequency field of frequency ω, a somewhat complex resonance phenomenon will occur when $\omega = \gamma B$. A detector measures the number of nuclei that go into resonance.

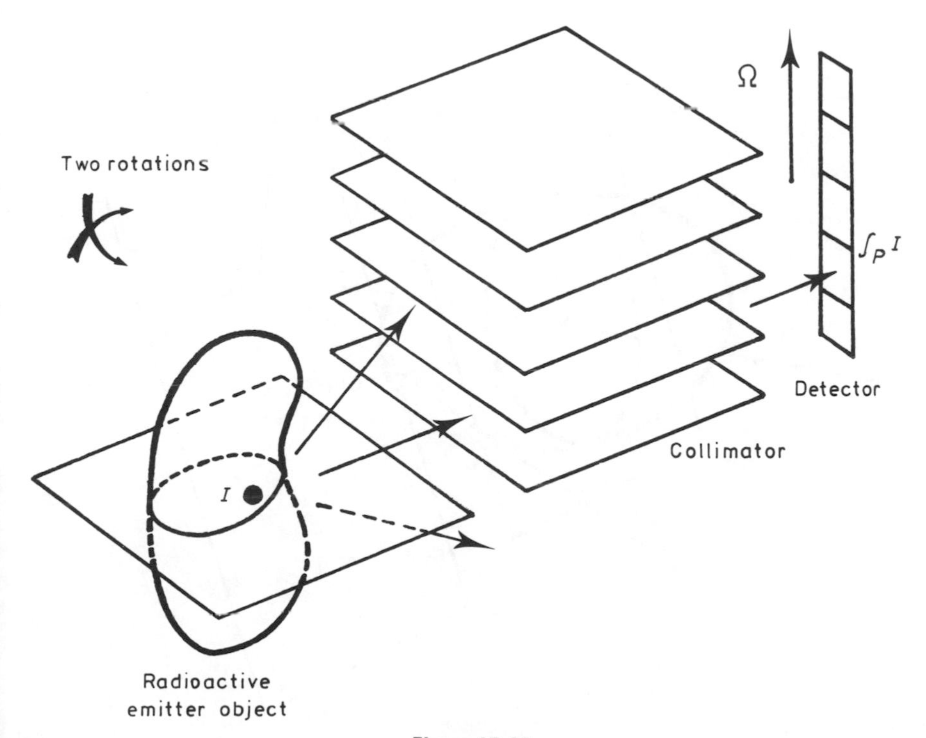

Figure 15.10

To excite the nuclei of a plane of the object, the following procedure is employed. The magnetic field B is carried by the vector $\mathbf{u}$ but its intensity varies linearly in a direction $\boldsymbol{\Omega} = \mathbf{grad}\ B$. Just one of the planes perpendicular to $\boldsymbol{\Omega}$ satisfies the resonance condition and the detector measures the integral of the object function over this plane.

In the preceding two examples, the integral over a plane of a function f characterizing the object is recorded. This is done for all the planes. More precisely, if the planes are rearranged as a family of parallel planes, the measurements provide, for each direction $\boldsymbol{\Omega}$, the projections

$$P_{\boldsymbol{\Omega}}(x') = \iint f(x', y', z')\, \mathrm{d}y'\, \mathrm{d}z'$$

where the integral is taken in the (x', y', z') space whose x' axis is along $\boldsymbol{\Omega}$ (Figure 15.11).

As in the two-dimensional case, the one-dimensional Fourier transform of

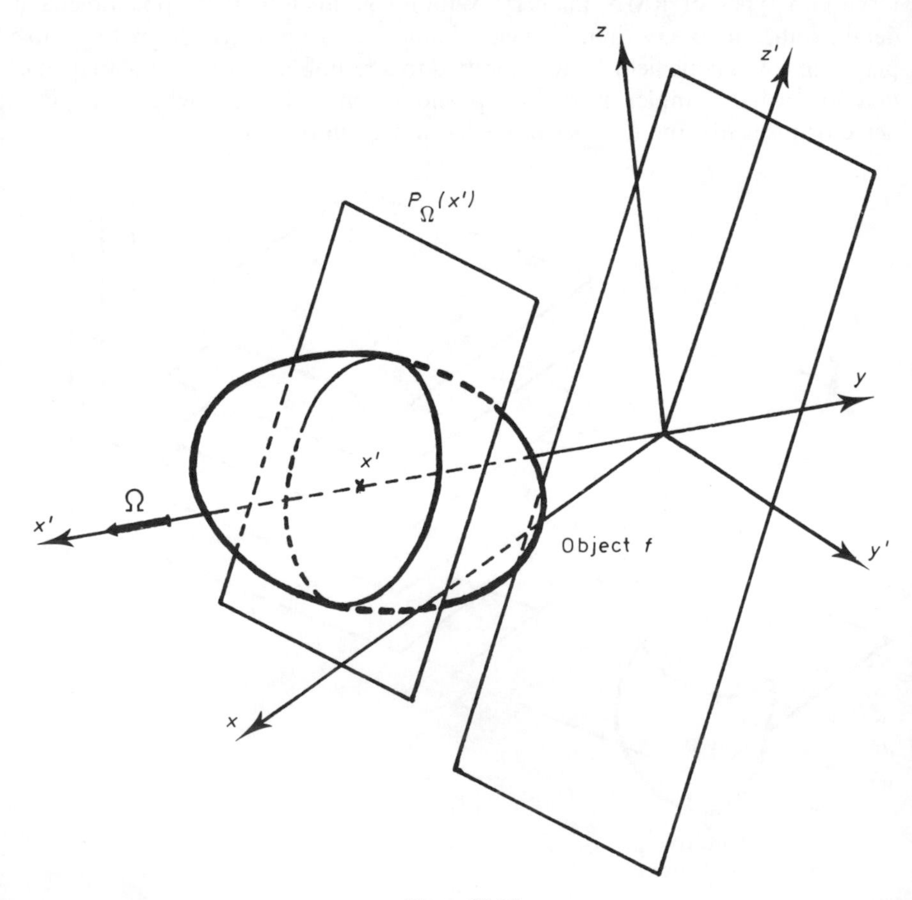

Figure 15.11

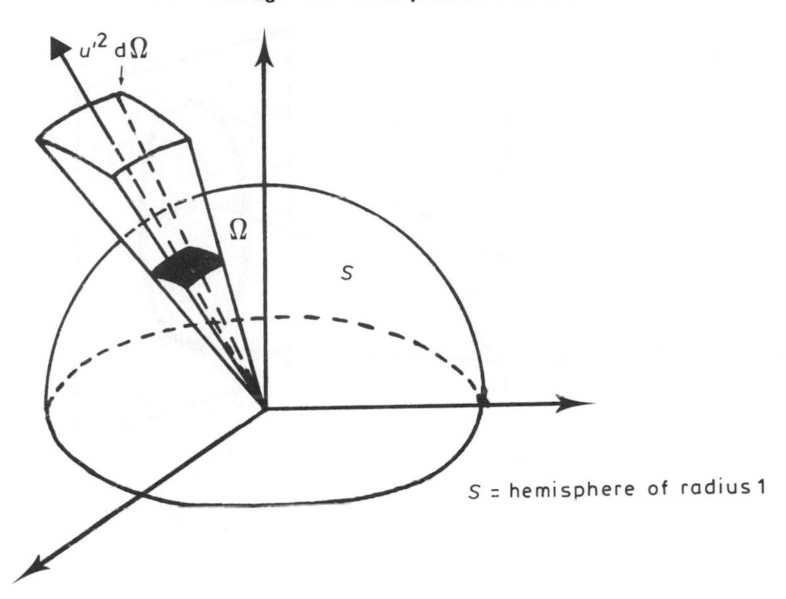

Figure 15.12

$P_\Omega(x')$ is equal to the restriction to the line in the $\mathbf{\Omega}$ direction of the three-dimensional Fourier transform of f. Thus, being given the projections in all directions of the space is enough to determine $\tilde{f}$ at every point and allows f to be calculated:

$$f(M) = \iint_S \int_{-\infty}^{\infty} u'^2 \tilde{P}_\Omega(u') e^{2\pi i u' x} \, du' \, d\Omega$$
$$= -\frac{1}{4\pi^2} \frac{d^2}{dx'^2} P_\Omega(x') = P^*_\Omega(x')$$

One therefore passes from knowledge of the projections P_Ω to f via the product of a second derivative and a retroprojection (Figure 15.12).

15.3 Codings with non-separated sections and not reducible to the three-dimensional Radon transform

Consider, to begin with, the case of non-transverse axial tomography. In absorption, the absorbing object is lit by a point source and no longer by a beam of parallel rays. When the object revolves about the z axis, mixing of the information coming from different planes of sections perpendicular to Oz takes place (Figure 15.13).

In imagery by code-breaking, the emitter object projects the shadow of a code on to the detector (Figure 15.14). One can reconstruct the image by a method of partial deconvolution without involving the Radon transform at all.

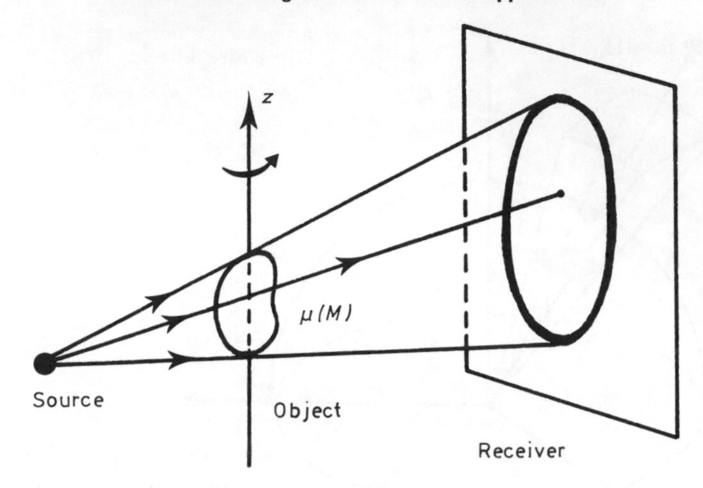

Figure 15.13

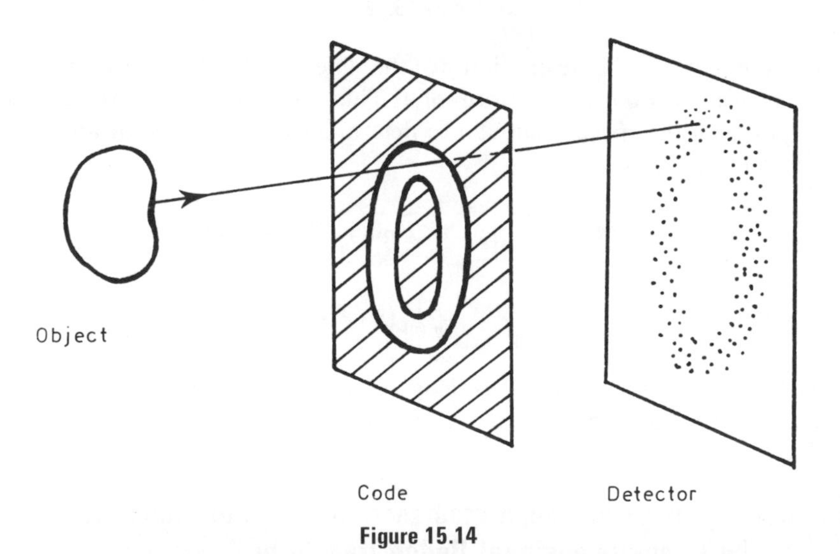

Figure 15.14

To obtain images of microplasmas, six laser beams are focused on a small glass ball containing a gas. Under the action of the energy received, a plasma is created in the gas and this emits various radiations. One is particularly interested in drawing up a chart of the X-ray emission. This chart provides information on the homogeneity of the plasma.

Figure 15.15 gives an idea of the somewhat cumbersome arrangement of lenses serving to focus the beams. Since the life of the plasma is of the order of only 10^{-10} s, all information must be obtained simultaneously. For this, the

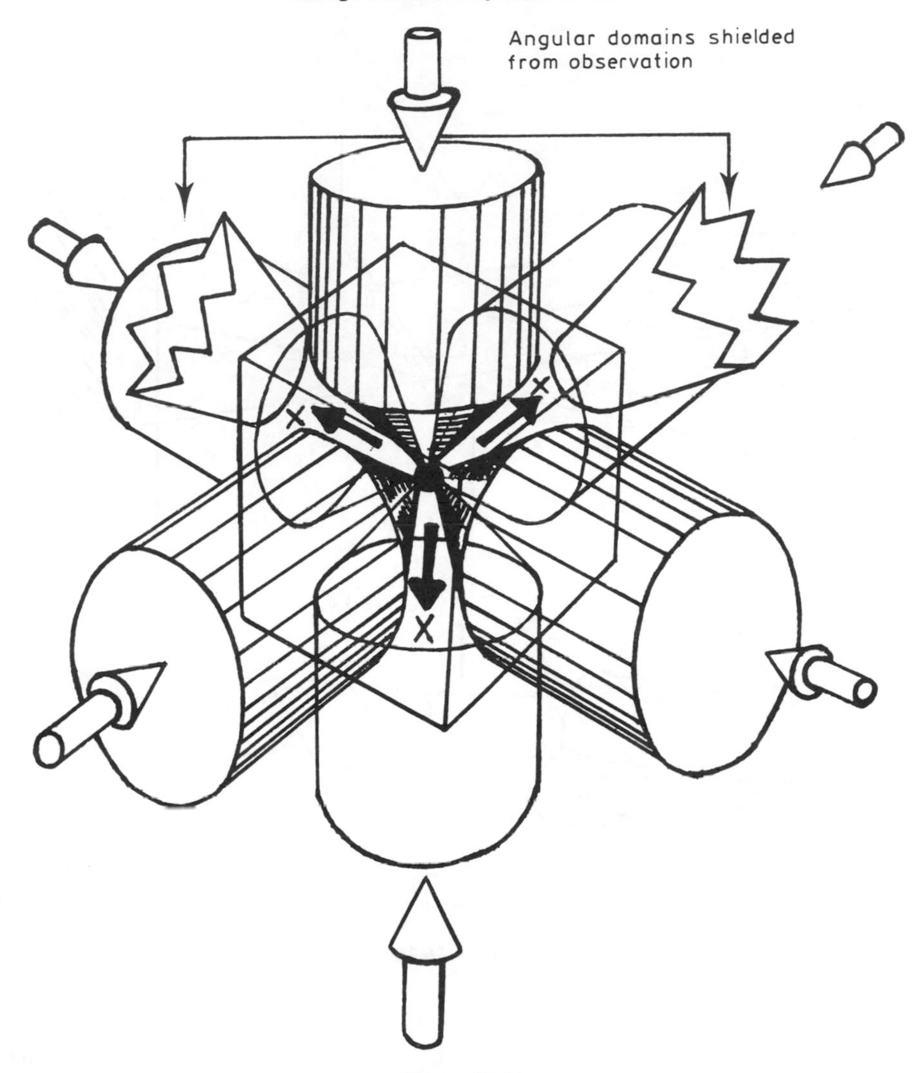

Figure 15.15

scheme of imagery by code-breaking is employed. The X-rays emitted by the plasma act on the receiver through a mask consisting of a series of slits.

Suppose, to begin with, that the mask consists of just one slit (Figure 15.16). At each point of the receiver, the flow received is the integral of $I(M)$ over the plane determined by the point and the slit. This flow is the same along a line parallel to the slit.

In view of the respective dimensions of the ball (0.1 mm) and the ball–mask distance (5 cm), everything happens just as if the planes P_1 and P_2 were parallel. One therefore measures the projections of the object on the direction $\mathbf{\Omega}$ perpendicular to the plane determined by O and the slit.

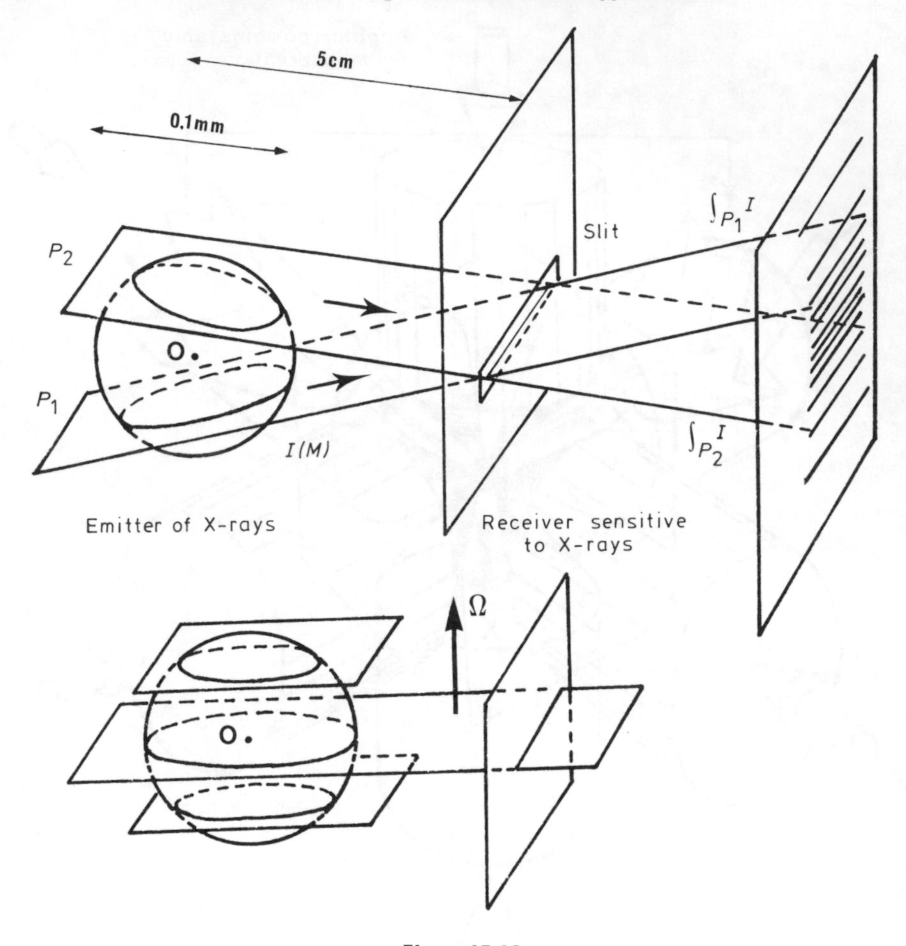

Figure 15.16

If the mask now consists of several slits, there is geometric separation of the groups of projections on the plane of the detector. Herein lies a fundamental difference from imagery by code-breaking: a mask comprising several slits is not equivalent to a single code but to several independent codes. This procedure yields the projections in various directions (Figure 15.17).

The nature of the problem depends on the data-collection geometry.

Consider first the case where the slits all pass through the same point on the mask. We see that the projections (in the sense of a three-dimensional Radon transform) of the object are nothing but the projections (in the sense of a two-dimensional Radon transform) of the object projected on to the mask (Figure 15.18). One thus comes down to the two-dimensional Radon transform. The third dimension is lost.

Another geometrical arrangement consists of taking slits tangent to a circle.

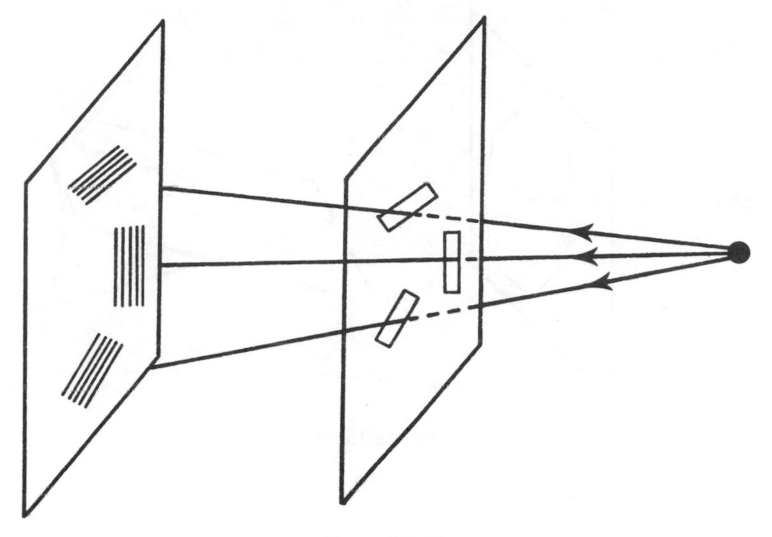

Figure 15.17

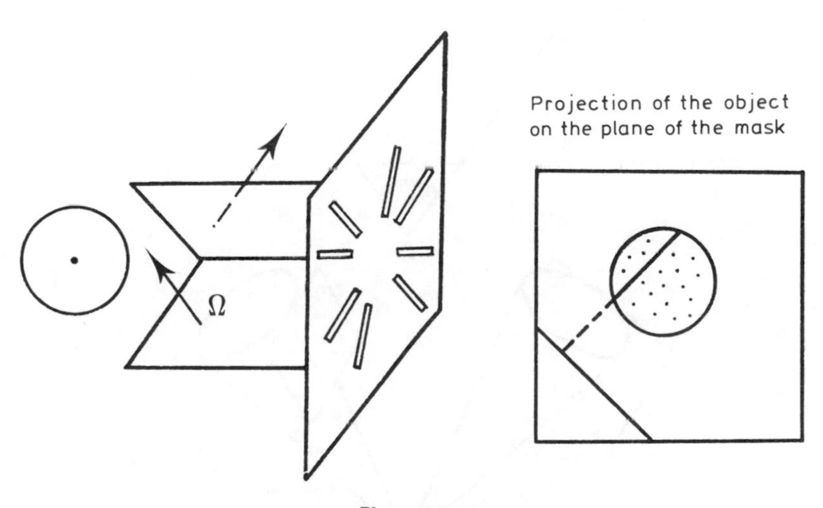

Figure 15.18

One then obtains the projections (in three dimensions) on directions of planes tangent to a cone (Figure 15.19).

One can also have a more complicated configuration: two masks at 70° with slits tangent to a circle (Figure 15.20). One obtains projections on directions of planes tangent to two cones.

We thus see that we are a long way from having all the necessary data for the reconstruction of the object with the help of the three-dimensional Radon

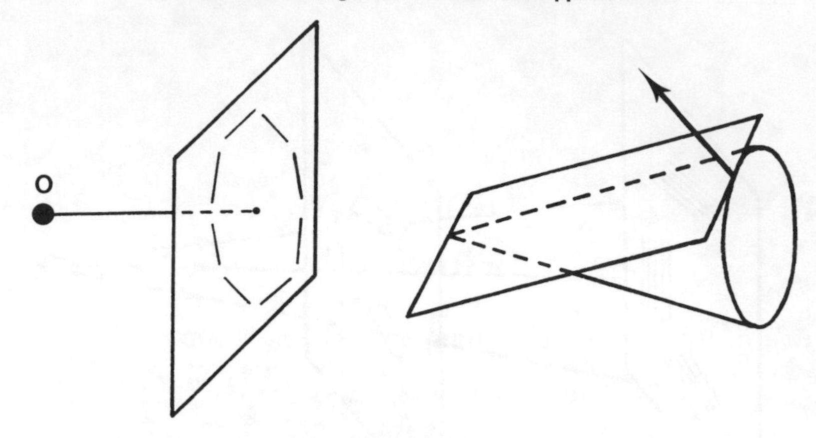

Figure 15.19

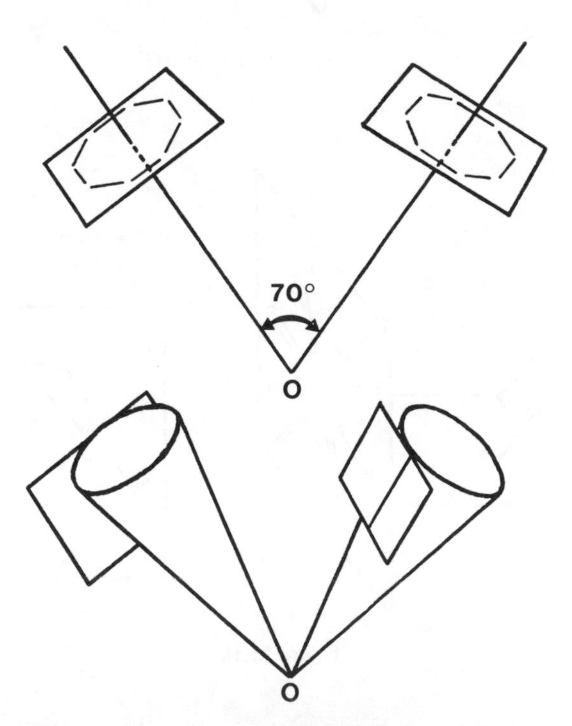

Figure 15.20

transform. This is due to the fact that one only has access to a limited angular zone around the object.

In this case it is necessary to employ a method analogous to that described in two dimensions. Studies in this are are in progress and have already produced encouraging results.

Chapter 16 Creation of fractal objects by diffusion

M. Rosso, B. Sapoval, J.-F. Gouyet and J. D. Colonna
Ecole Polytechnique (Palaiseau)

16.1 Introduction

At the atomic level, diffusion in a solid can be described as the result of the random walk of particles in a host network. If the parameters characterizing the advance of the diffusing substance, regarded as a continuum, obey the classical laws of diffusion, one can, on the other hand, expect, at the atomic level, the structures or objects created to be much more complex. The numerical simulation work described in this article has such structures as its object of study. For two-dimensional systems, two important results have been obtained (Sapoval *et al.*, 1985a): the 'diffusion front' is a fractal object of dimension 1.75; it may be identified with the external envelope of the percolation mass (Voss, 1984).

The different steps of the diffusion process obtained by means of this simulation can be visualized, and such a visualization is the purpose of a video film (Sapoval *et al.*, 1985b). The real interest in the visualization was twofold:

(a) To display the internal similarity (Mandelbrot, 1982) of objects created by diffusion.
(b) To exhibit the erratic character of the time evolution of such objects. One sees that objects of semi-macroscopic size fluctuate in a random way, while the macroscopic parameters characterizing the type of the diffusion obey the usual laws of diffusion.

The principles underlying our simulation calculations have been presented in detail elsewhere (Sapoval *et al.*, 1985a; Rosso *et al.*, 1985). Let us recall the essentials: on a two-dimensional network (we deal here principally with the case of the square lattice), particles can jump at random from a site to one of its immediate neighbours, provided this latter is empty. Under these conditions one can show (Dieterich *et al.*, 1980) that the concentration $p(x, t)$ of the particles is the solution of the classical Fick equations for diffusion. If, further, one maintains a source of diffusing particles at constant concentration $p(0, t) = 1$ at

$x = 0$, $p(x, t)$ is given by

$$p(x) = \operatorname{erfc}\left(\frac{x}{l_D}\right) \tag{1}$$

where l_D, the diffusion length, is equal to $l_D = 2(Dt)^{1/2}$. D is the coefficient of diffusion of the particles on the network under consideration.

Two simulation techniques have been used:

(a) An initial technique, 'true' simulation of diffusion, follows the diffusion step by step, causing each particle to advance at random through the lattice, which initially is either empty or already partially occupied by the diffusion. This first method is very extravagant of computer time.

(b) A second method is based on the fact that we know the distribution $p(x, t)$ of the particles. At a given moment in time t, the particles are distributed at random over the lattice, in such a way that their mean concentration $\langle p(x, t) \rangle$ satisfies the relation (1). What we have here is a picture of a configuration of particles at a given moment.

If the two methods result in the same distribution of concentration $p(x, t)$, the second is plainly much shorter than the true simulation, but ignores the possible effect of correlations between particles on the evolution of the diffusion. However, we estimate that this effect is negligible as far as the properties that interest us are concerned, in particular the geometry of the object obtained by diffusion. On the other hand, the true simulation is essential for getting to know the dynamic behaviour of this object.

16.2 Internal similarity and fractal properties of the diffusion front

We give the name 'diffusion front' to the boundary of the set of diffused particles still connected to the source of the diffusion. This definition rests on that of connections between particles: in our square lattice we consider two particles to be connected if they occupy immediately neighbouring sites.

The physical role of the diffusion front can be illustrated by means of the following example. If the diffusing atoms are metallic and the host network is insulating, and if one supposes that there is only metallic contact between two immediately neighbouring sites, then none of the diffused particles remain in metallic contact with the source. This is shown in Figure 16.1: the occupied sites are shown in white or grey. Amongst these, one distinguishes the mass of sites isolated from the source (the top line of the figure) and a set of sites connected to the source. The white sites represent the boundary of this set: in our electrical picture, this is the boundary of the equipotential region at the source.

Figure 16.1 Schematic representation of the system of diffused particles. The occupied sites are shown in white or grey. Amongst these, one distinguishes the mass of sites isolated from the source (the top line of the figure) and a set of sites connected to the source. The white sites represent the boundary of this set

We have chosen to describe these objects in easier to follow 'geographical' terms. The region connected to the source is called the 'land'. The empty region, not reached by the diffusion, constitutes the 'sea'. The strip of land moistened by the sea is naturally the 'beach' (in white, here), whilst one also distinguishes the 'islands' (occupied sites, isolated from the land) and the 'lakes' (empty sites, isolated from the sea).

Let us now look at the properties of the diffusion front, our 'beach'. We have shown that, for long diffusion times, it is a fractal object of dimension $d_H \sim \frac{7}{4} = 1.75$. A 'landscape' obtained for a relatively long diffusion time is represented in Figure 16.2. It relates to a 256×256 point picture. The diffusion length is $l_D = 300a$, where a represents the inter-site distance; it is attained for a time $t = 2.25 \times 10^4\ a^2/D$.

The mean distance x_f from the frontier (the beach, shown here in white) to the source is about $114a$ and its width σ_f of the order of $12a$. There are 2944 particles altogether.

See how very convoluted the beach appears to be, and it is in fact only a *line*: this is due to the high dimension (1.75) of this object. Figure 16.3 shows a diffusion front obtained for $l_D = 51\,200$: only the front itself is represented here, in white. The internal similarity of the object is shown clearly here. Indeed, these four pictures have very similar features, although they represent four different enlargements of the diffusion front (from $\times 1$ at upper left to $\times 8$ at lower right).

We have also shown (Sapoval *et al.*, 1985a; Rosso *et al.*, 1985) the similarity

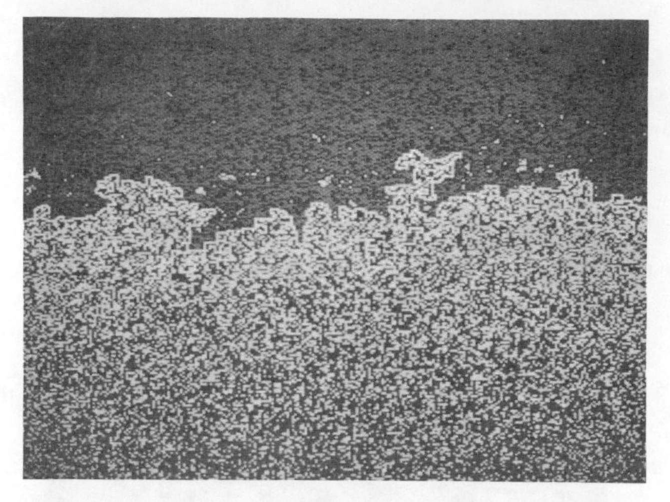

Figure 16.2 Representation of the system of diffused particles for a time of the order of $2.25 \times 10^4 a^2/D$

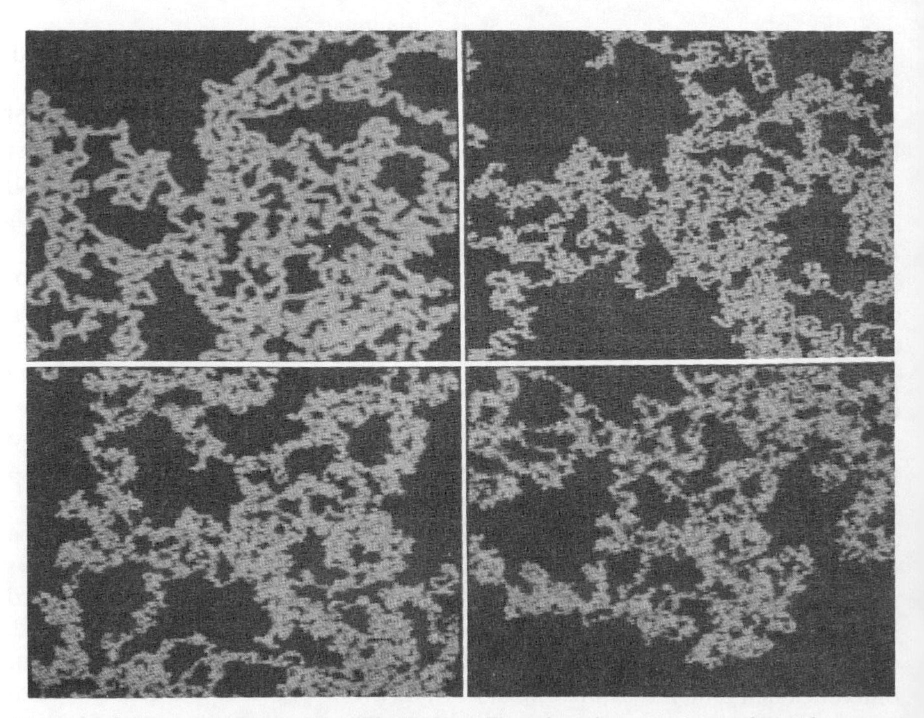

Figure 16.3 Internal similarity of the diffusion front. These four pictures represent four enlargements of the diffusion front from $\times$ 1 at upper left to $\times$ 8 at lower right

that there is between the diffusion front, in the limit when $l_D \to \infty$, and the external envelope of the percolation mass (Voss, 1984). From this property together with the fractal nature of the diffusion front we have been able to deduce the following relationships:

$$x_f = k_x(Dt)^{1/2} \tag{2}$$

$$\sigma_f = k_\sigma(Dt)^{\alpha_\sigma/2} \qquad \text{with } \alpha_\sigma = \nu/(1+\nu) = 4/7 \tag{3}$$

$$N_f = k_N L(Dt)^{\alpha_N/2} \qquad \text{with } \alpha_N = 1/(1+\nu) = 3/7;\ L \text{ is the sample length} \tag{4}$$

$$d_H = 1 + \frac{1}{\nu} = \frac{7}{4} = 1.75 \tag{5}$$

where ν is the critical exponent of the correlation length ξ of the classical percolation.

In the square lattice the constants k_x, k_σ and k_N are respectively equal to 0.856, 0.68 and 1.34.

16.3 Dynamic behaviour of the diffusion front

The second crucial result shown up by the visualization concerns the dynamic evolution of the diffusion front. To study this behaviour, our simulation starts from a distribution of particles determined by the second procedure (described in Section 16.1), such as that shown in Figure 16.2. The diffusion of this distribution of particles is then simulated by the first procedure ('true' simulation).

Starting out from the initial distribution of Figure 16.2, time is allowed to pass as a sequence of elementary events occur. The following are regarded as 'elementary events' of the simulation: the random choice of a particle and then one of four possible jumps to an immediately neighbouring site. The jump will then actually take place or not according as to whether this site is empty or occupied. By looking at the system periodically, we make a study of the dynamics by sampling, all k elementary events being considered.

It is useful here to go over again, in greater detail, some of the results presented in Section 16.2. We have shown (Sapoval *et al.*, 1985a; Rosso *et al.*, 1985) that a distribution of particles taken at a given moment can be studied in the setting of the theory of percolation in a situation where the concentration is not constant. This is what we call 'percolation in a gradient' (Rosso *et al.*, 1985; Sapoval *et al.*, 1986a). In this picture, the masses (here, the islands or the lakes) have a characteristic size ξ which depends on their position. When x tends to x_f it increases to a maximal value of the order of σ_f. The exterior boundaries of the islands and the lakes have the same fractal geometry as the beach but only up to the ξ-level (Sapoval *et al.*, 1985a).

If one looks more closely at a 'lake near the beach' or an 'island near the beach' (Figures 16.4 and 16.5), one sees that the status of such objects is very

Figure 16.4 A 'lake (in grey) near the beach (in white)'

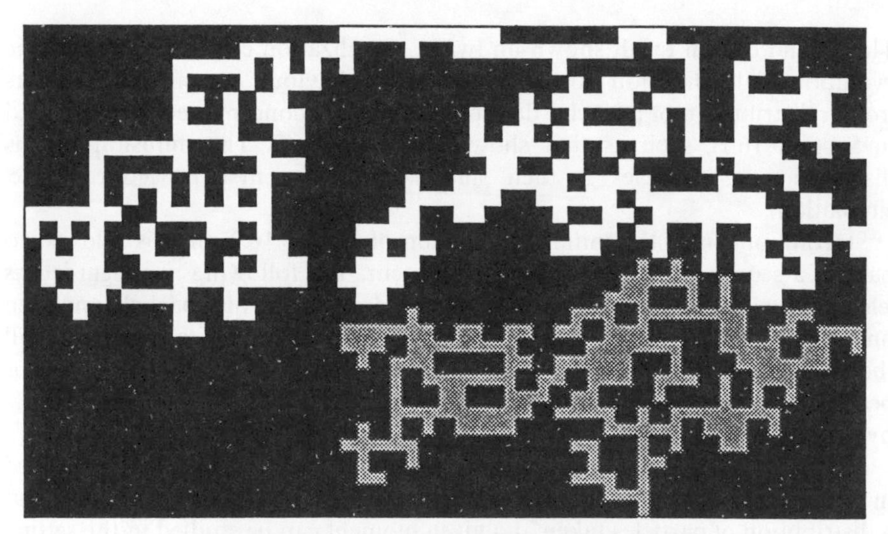

Figure 16.5 An 'island (in grey) near the beach (in white)'

precarious. In fact, the movement of a single particle is enough to transform an island into a peninsula or a lake into a bay. This means that, during the diffusion, there are microscopic events (movement of a particle) that instigate semi-macroscopic changes of size (size of an island or lake) in the frontier. Fairly long periods of time during which nothing important happens are thus interrupted by various remarkable events.

From our simulation of diffusion, we have taken one of these remarkable

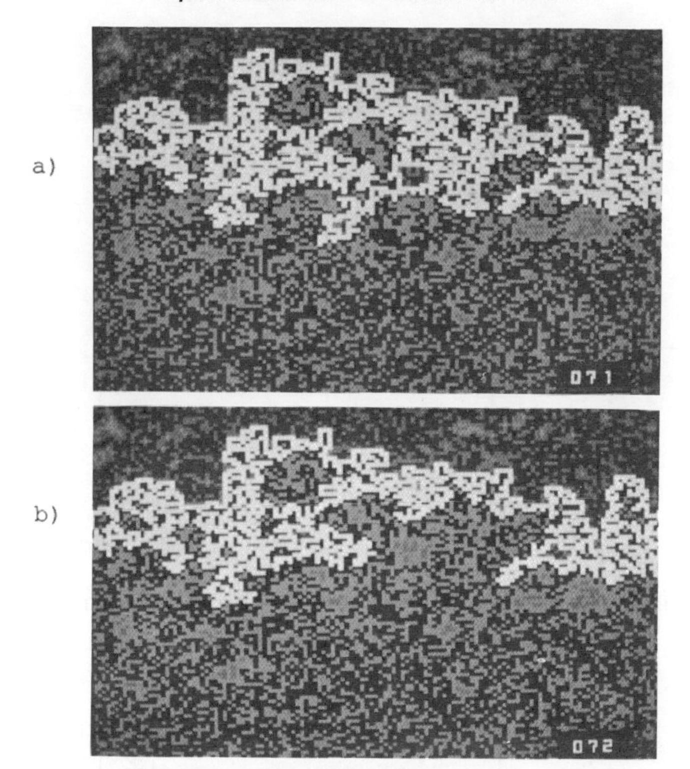

Figure 16.6 Representation of the system of diffused particles at the instants (a) $n = 71$ and (b) $n = 72$: a peninsula is transformed into an island

events, shown in Figure 16.6. These views have been selected from a sequence of a hundred successive states of the diffusion front, obtained for intervals of time

$$t_n = t_0 + n\Delta t$$

where

$n = 1$ to 100

$t_c = 2.25 \times 10^4\, a^2/D$

$\Delta t = 40$ elementary events

The time corresponding to an elementary event is here equal to $0.6 \times 10^{-5}\, a^2/D$.

Figures 16.6(a) and (b) shows the remarkable event consisting of the transformation of a peninsula into an island (the number of points on the beach decreases). The time evolution of the number N_f of points of the diffusion front is represented in Figure 16.7 for $n = 50$ to 100. The succession of calm periods and catastrophic events can be clearly seen here.

A preliminary quantitative study of the dynamics of the diffusion front

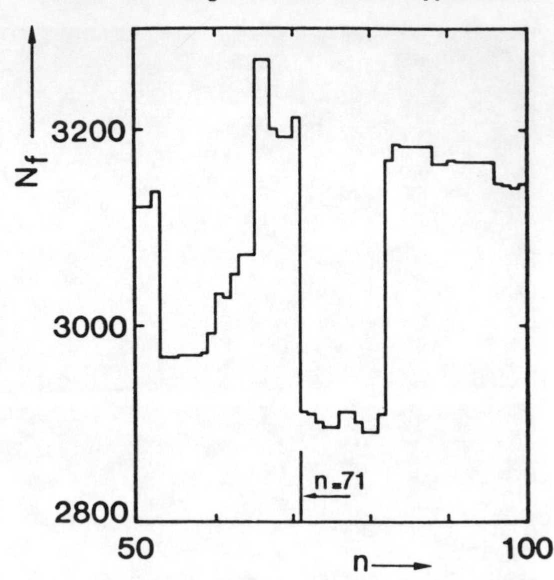

Figure 16.7 Representation of the time evolution of the number N_f of points of the diffusion front

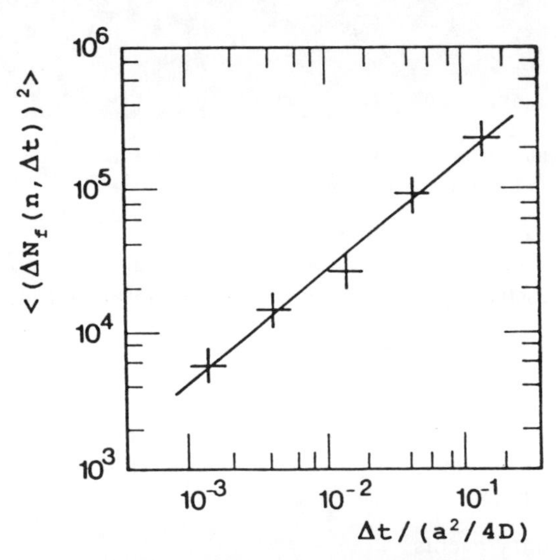

Figure 16.8 Variation of the mean value of the increase in the number of points of the front, $\Delta N_f(n, \Delta t)$, as a function of the sampling time Δt

(Sapoval *et al.*, 1986a) shows that the increase in the number of points of this front

$$\Delta N_f(n, \Delta t) = N_f(t_n) - N_f(t_{n-1})$$

obeys a law, namely:

$$\langle [\Delta N_f(n, \Delta t)]^2 \rangle \propto \Delta t^{0.8} \qquad \text{for } \Delta t < a^2/4D \tag{6}$$

This power law, illustrated in Figure 16.8, is characteristic of an antipersistent fractional brownian motion (Mandelbrot and Wallis, 1968).

We have predicted that this erratic dynamic behaviour of the diffusion front could have practical consquences in the physical systems close to those described by our two-dimensional model, for example layered intercalation compounds (Sapoval *et al.*, 1986b). We have thus predicted the existence of an intercalation noise of $f^{1.8}$ (Sapoval *et al.*, 1986a).

For sampling times $\Delta t > a^2/4D$, one observes a stabilization of the mean fluctuations $\Delta N_f(n, \Delta t)$, which then corresponds to a noise of $1/f$.

16.4 Conclusion

Simulation of diffusion in a bidimensional lattice has shown that at the atomic level the diffusion front has a fractal geometry; its dynamic evolution is erratic and one observes semi-macroscopic fluctuations over ultra-short time scales, corresponding to the jump of a single particle of the diffusion front. Our results, depending as much on the geometry of the front as on its dynamics, should have very interesting consequences in bidimensional systems such as those produced by intercalation in layered compounds (Sapoval *et al.*, 1986a). The disorderliness at every level is to be found equally in three-dimensional diffusion fronts (Gouyet *et al.*, 1986) whose study is somewhat more complicated.

References

Dieterich W., Fulde P. and Peschel I. (1980) Theoretical models for superionic conductors, *Adv. Phys.*, **29**, 527.

Gouyet J.-F., Rosso M and Sapoval B. (1986) Percolation in a concentration gradient, in L. Pietronero and E. Tosatti (Eds.), *Fractals in Physics*, Elsevier Science Publishers B.V.

Mandelbrot B. (1982) *The Fractal Geometry of Nature*, Freeman, San Francisco.

Mandelbrot B. and Wallis J. W. (1968) Fractional brownian motions, fractional noises, and applications, *SIAM Rev.* **10**, 422.

Rosso M., Gouyet J.-F. and Sapoval B. (1985) Determination of percolation probability from the use of a concentration gradient, *Phys. Rev.*, **B32**, 6053.

Sapoval B., Rosso M. and Gouyet J.-F. (1985a) The fractal nature of a diffusion front and the relation to percolation, *J. Phys. Lett.*, **46**, L149.

Sapoval B., Rosso M, Gouyet J.-F., Colonna J. F. and Boutin A. (1985b) Simulation of a diffusion process in a solid (broadcast by Imagiciel).

Sapoval B., Rosso M., Gouyet J.-F. and Colonna J. F. (1986a) Dynamics of the creation of fractal objects by diffusion and $1/f$ noise, *Solid State Ionics*, **18, 19**, 21.

Sapoval, B., Rosso M and Gouyet J.-F. (1986b) Simulation of fractal objects obtained by intercalation in layered compounds, *Solid State Ionics*, **18, 19**, 232.

Voss R. F. (1984) The fractal dimension of percolation cluster hull, *J. Phys.*, **A17**, L373.

Chapter 17 Irreversibility and the arrow of time

J. Chanu
Laboratoire de Thermodynamique des Milieux Ioniques et Biologiques, University of Paris VII

Traditionally, the kind of irreversibility that manifests itself at the level of macroscopic transport phenomena would be explained at the basic elementary level by means of time-reversible laws (Onsager's microscopic reversibility). However, such a point of view seems more and more difficult to reconcile with the irreversible behaviour of matter beyond the linear domain (i.e. beyond that branch of thermodynamics which includes heat conduction, isothermal and thermal diffusion, newtonian viscosity, etc.). The introduction of Benard's instability into a fluid layer and chemical and biochemical reactions which oscillate in both time and space are some of the non-linear phenomena characterized by a rupture of symmetry in time and space.

In the light of the work of I. Prigogine and his collaborators, we today know that such a spatio-temporal rupture of symmetry originates in the very heart of matter and lies at the base of constructive processes called *dissipative structures*. In another connection, irreversibility seems to play a fundamental motive role in the evolution of the universe (it should intervene in processes as fundamental as the gravitational collapse of stars).

The first problem to arise was that of establishing connections between the dynamical description of systems on the one hand and the properties that correspond to the increase of entropy as a function of time (the arrow of time), on the other hand. In a dynamical system, the sense in which time unfolds matters little. The general mathematical solution is as appropriate to future times ($t > 0$) as to past epochs ($t < 0$). Thermodynamics itself only deals with the unfolding of events in the future ($t > 0$). Prigogine then showed that dynamic evolution takes place via the application of a unitary operator, which leads to a dynamical group. On the other hand, in a time-directed evolution for which, moreover, $t > 0$ and $t < 0$ are mutually exclusive, the properties are those of a semi-group (non-unitary operator). Only one of the evolutions gives rise to physically realizable states. In view of this fact, the second principle takes on the appearance of a selection principle.

The passage from a dynamical description (dynamical group) to a semi-group only takes place through the intermediary of a transformation Λ breaking the symmetry. However, this only acts if, in fact, the initial conditions reveal a sufficiently unstable state (keen sensitivity to initial conditions). What is really involved is a probabilistic process which introduces a delocalization in space.

One of the most spectacular consequences of these new developments due to Prigogine is the discovery of the existence of a time T 'internal' to the evolving matter, a function of the age of its different component parts. This time is thus quite distinct from the time t given by astronomical clocks: it is no longer a parameter but actually an operator which can be used to justify the concept of temporal non-locality and thence the somewhat unexpected idea of 'duration' of the present.

The development of this theory, the framework of which is already well established in the context of the irreversibility of dynamical systems, has really only just begun. Other areas that touch on macroscopic physics (turbulence, for example) or even biology should also be tackled.

Bibliography

Mandelbrot B., Cherbit G., Chanu J. and Le Méhauté A. (1985) Le concept de dimension non entière en physique. L'entropie et la flèche du temps. Séminaires Interdisciplinaires Hausdorff, Laboratoire de Thermodynamique des Milieux Ioniques et Biologiques, University of Paris VII.

Prigogine I. (3rd ed. in press 1986) *Physique, Temps et Devenir*, French edition in the hands of J. Chanu, Masson Ed., Paris.

Chapter 18 Thermodynamic entropy and information

M. Courbage
Laboratoire de Probabilités, P. and M. Curie University

In this short article we set out the link that exists between the thermodynamic entropy as a state function which increases monotonically with time and the mathematical theory of information. With the help of the concepts of this theory one can express the second principle as the impossibility of fitting states to infinite information with regard to equilibrium.

18.1 Information and thermodynamics

The existence of connections between the second principle of thermodynamics and the statistical concepts of information is an intuitive idea of widespread currency among physicists. In fact, this idea goes back to the work of Boltzmann who introduced the famous entropy formula

$$-\int \mathrm{d}r\,\mathrm{d}v\, f(r,v,t)\log f(r,v,t) \tag{1}$$

where $f(r,v,t)$ is the probability distribution of a gas interacting through collisions between pairs of molecules. Boltzmann later interpreted this formula as the probability measure of the distribution f (regarding this subject, see the excellent article of M. J. Klein [10]). Thanks to this formula, Boltzmann was able to give his theorem H the well-known statistical meaning according to which the distribution of a system evolves in time in accordance with the direction of the increase of its probability.

The concept of 'probability measure of a state of the system', subsequently the measure of its disorderliness, rested on an abstract concept until the work of the founders of the mathematical theory of communication, particularly that of Shanon and McMillan. This work was concerned with characterizing the degree of uncertainty of a random experiment α having k possible outcomes $(\alpha_1,\ldots,\alpha_k)$ with probabilities $p_1,\ldots,p_k$. Shanon concluded that the formula:

$$H(\alpha)=H(p_1,\ldots,p_k)=-\sum_{i=1}^{k} p_i \log p_i \tag{2}$$

can be taken as a mean measure of the difficulty of predicting the result of a draw and called this quantity the *entropy of the experiment* α because of its interesting properties, which are given later on in this article. (For further information on this subject the reader is referred to the works of Brillouin [2], Khinchin [9] and Yaglom and Yaglom [17]. Apart from the analogy of this formula with the well-known formula: $S = -K \log P$, which reduces to (2) on using Stirling's formula, it seems to us that, up to now, there has been no theory that enables one to establish a connection between thermodynamics and information theory.

Boltzmann's idea of associating the entropy of an isolated system with the measure of its disorder can nowadays take on a more precise meaning thanks to new notions in the theory of dynamical systems. In particular, Kolmogorov's concept of entropy of a dynamical system enables one to characterize globally the complexity of this system for a stationary probability distribution. This quantity, linked, in another connection, with the fractal dimension (there is an extensive literature on the subject of this link), should be clearly distinguished from Boltzmann's thermodynamic entropy defined by (1).

The global complexity of dynamical systems thus leads to the problem of irreversibility being posed in these terms: in what type of dynamical systems can one expect the second principle to hold?

It should be remembered here that the second principle is not only concerned with the increase of entropy but equally with various impossibilities, among them the impossibility of spontaneous transformation of heat into work, etc. Formulated in a microscopic setting, this is expressed by the non-invariance of the set of physical states of a system under the inversion of time (which, in a classical system, is the operation of inversion of speeds). Breaking of temporal symmetry and increase in disorder should thus derive *in an intrinsic way* from the properties of the dynamical system.

18.2 Distance—information from one probability distribution with respect to another

To begin with, recall the significance of formula (2). It is clear that predicting the result of tossing a coin is more difficult if the two outcomes are equiprobable than when one of the outcomes is more probable than the other. A quantity $H(p_1, \ldots, p_n)$ which measures the degree of uncertainty of a random experiment should possess a certain number of essential properties:

(a) The degree of uncertainty of random experiments with n possible outcomes is maximum when the outcomes are equi-probable.

(b) The degree of uncertainty of two successive experiments α and β of joint probabilities $p(\alpha_i, \beta_j)$ is equal to the uncertainty of α plus the remaining uncertainty of β when α is known:

$$H(\alpha \vee \beta) = H(\alpha) + H(\beta/\alpha)$$

Here, $H(\beta/\alpha)$ is the conditional entropy defined by

$$H(\beta/\alpha) = \sum_{i=1}^{n} p(\alpha_i) H(p(\beta_1/\alpha_i), \ldots, p(\beta_m/\alpha_i)) \tag{3}$$

where

$$p(\beta_j/\alpha_i) = \frac{p(\beta_j, \alpha_i)}{\rho(\alpha_i)} \tag{4}$$

(c) The degree of uncertainty does not change if one adds to the experiment of probability zero:

$$H(p_1, \ldots, p_n, 0) = H(p_1, \ldots, p_n)$$

One can then prove the following theorem:

Theorem (Khinchin [9]) Every continuous symmetric function of $p_1, \ldots, p_n$, with $p_i \geqslant 0, \sum_{i=1}^{n} p_i = 1$, satisfying properties (a), (b) and (c) is of the form:

$$H(p_1, \ldots, p_n) = -\lambda \sum_{i=1}^{n} p_i \log p_i$$

It follows from this that the major contributions to H come from events that are neither too likely nor too unlikely. (Recall that the function $-p \log p$ has the form shown in Figure 18.1, where log means $\log_2$. It attains its maximum for $p = \frac{1}{2}$). One can equally interpret the formula (2) in the following way.

An uncertainty function associated with an event of probability $p, p \to I(p)$ is a decreasing continuous real-valued function such that the uncertainty of two successive events α_i and β_j is equal to the sum of the uncertainty of α_i and the uncertainty of β_j given α_i:

$$I(p(\alpha_i, \beta_j)) = I(p(\alpha_i)) + I(p(\beta_j/\alpha_i)) \tag{5}$$

As $p(\alpha_i, \beta_j) = p(\alpha_i) p(\beta_j/\alpha_i)$ it follows from this that I must satisfy the condition: $I(xy) = I(x) + I(y)$. Moreover, such a function can only be of the form

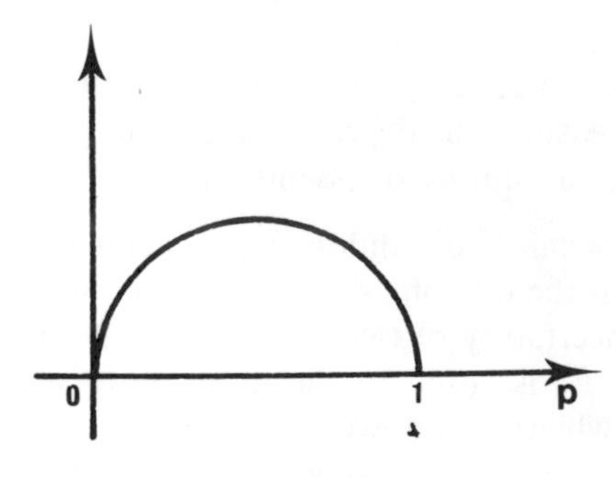

Figure 18.1

$I(p) = -\log p$. Shanon's formula (2) is thus the mean uncertainty of the random experiment α.

The problem of statistical mechanics outside equilibrium consists of the study of the evolution of non-stationary states towards the equilibrium state. Onc then needs to express the increase in uncertainty on passing from a distribution $p = (p(\alpha_i) = p_i)$ to a distribution $q = (q(\alpha_i) = q_i)$. This is equal to

$$(-\log q_i) - (-\log p_i) = \log p_i/q_i, \qquad \text{for the event } \alpha_i$$

The mean of this function with respect to the distribution p_i:

$$\Omega_q(p, \alpha) = \Omega_q(p) = \sum_{i=1}^{n} p_i \log p_i/q_i \tag{6}$$

represents the gain of information of the distribution (p_i) relative to q_i.* One can easily check its principal properties:

(a) $\Omega_q(p)$ is positive and attains its zero minimum if and only if $p_i = q_i, \forall_i$.
(b) $\Omega_q(p) = +\infty$ if p_i assigns a positive probability to an event α_i of probability $q(\alpha_i) = 0$. By (6), the passage from p_i to q_i is accompanied by an infinite increase in the uncertainty and consequently p contains infinite information with respect to q. When $\Omega_q(p) = +\infty$, we say that p is not comparable with q.
(c) Among all the distributions p comparable with $q, \Omega_q(p)$ is maximum for a certain determinist distribution, i.e. a distribution where just one event α_i has probability 1.
(d) The information of a compound experiment $\alpha \vee \beta = \{(\alpha_i, \beta_j)\}$ is equal to thc sum of the information rclative to α and the information relative to β if α is known (see (4)):

$$\Omega_q(p, \alpha \vee \beta) = \Omega_q(p, \alpha) + \Omega_q(p, \beta/\alpha) \tag{7}$$

$$\Omega_q(p, \beta/\alpha) = \sum_i p(\alpha_i)\Omega_q(p, \beta/\alpha_i) \tag{8}$$

$$\Omega_q(p, \beta/\alpha_i) = \sum_j p(\beta_j/\alpha_i) \log \frac{p(\beta_j/\alpha_i)}{q(\beta_j/\alpha_i)} \tag{9}$$

Having set out these preliminary notions concerning the increase in uncertainty under passage from one distribution to another,† let us examine their connection with thermodynamic entropy.

Consider a conservative dynamic flow S_t on the phase space Γ of the system, μ being the 'microcanonical' measure invariant under the flow. Following Gibbs, we describe a state of the system by means of a probability distribution $\rho(\omega) \geqslant 0$ such that $\int_\Gamma \rho_t(\omega)\, d\mu(\omega) = 1$. We know that $\rho_t(\omega)$ is a solution of the Liouville equation and then, because of the invariance of the measure μ, by Liouville's

*By definition, the increase in uncertainty is equal to the loss of information.
†Such a functional has already been used in statistics and in the theory of information transmission by Kulback [11], Perez [15] and several Soviet authors [6, 7].

theorem, the quantity:

$$\Omega(\rho_t) = \int_\Gamma \rho_t(\omega) \log \rho_t(\omega) \, d\mu(\omega)$$

remains constant. It is in fact well known that classical dynamics is reversible and that an intrinsic characterization of the irreversibility should follow from a detailed analysis of the phase space of the system, or rather of the sequence of observable events, to be precise.

Indeed, one can associate two classes of observables with a dynamical system: the future observables and those of the past. To an observation of finite precision, there corresponds a partition $\mathscr{P}$ of the phase space into k disjoint regions of finite measure, $P_0, P_1, \ldots, P_{k-1}$. Supposing that at the instant $t=0$ the system is at a point ω of Γ, one considers the series of such observations made at regular intervals of time $0, \tau, 2\tau, \ldots, n\tau, \ldots$. To the dynamic flow $\omega \to S_t\omega = \omega_t$ corresponds the interation $T^n\omega = (S_\tau)^n\omega$ where $T = S_\tau$. A series of future observations of the system is represented by the sequence $u_n(\omega), n \geqslant 0$ where $u_n(\omega)$ is the number of the region where the system happens to be at the instant $n\tau$, *i.e.* $T^n\omega \in P_{u_n(\omega)}$. Another series of past observations can be introduced in a similar way for $n \leqslant 0$. An event of the form $P_{u_0} \cap T^{-1}P_{u_1} \cap \cdots \cap T^{-n}P_{u_n}$ thus corresponds to the set of phase points ω such that $T^i\omega \in P_{u_i}$, $i = 0, 1, \ldots, n$. The set of these elementary events constitutes a partition of the space Γ which will be written $\mathscr{P} \vee T^{-1}\mathscr{P} \vee \cdots \vee T^{-n}\mathscr{P}$ or, more simply, just $\mathscr{P}_n$. It now seems natural to define the entropy of a non-stationary distribution ν with respect to a given equilibrium distribution μ by means of the relative information of ν with respect μ for the future observations, i.e.

$$\Omega_\mu(\nu, \mathscr{P}) = \lim_{n \to \infty} \Omega_\mu(\nu, \mathscr{P}_n) \tag{10}$$

where, according to definition (6),

$$\Omega_\mu(\nu, \mathscr{P}_n) = \sum_{u_0, \ldots, u_n} \nu(P_{u_0} \cap \cdots \cap T^{-n}P_{u_n}) \log \frac{\nu(P_{u_0} \cap \cdots \cap T^{-n}P_{u_n})}{\mu(P_{u_0} \cap \cdots \cap T^{-n}P_{u_n})} \tag{11}$$

One of the major problems of statistical mechanics outside equilibrium is to find a characterization of the physical properties of the statistical sets (in Gibbs' sense) describing the state of the system outside equilibrium (apropos of this, see Penrose's discussion [14]). The functional $\Omega_\mu(\nu, \mathscr{P})$ furnishes one possible candidate and the remainder of this article is devoted to this matter.

Consider observations associated with $\mathscr{P}$ having the following two properties:

(a) $\vee_{-\infty}^{+\infty} T^i\mathscr{P}$ is the partition of Γ into individual points (i.e. almost every point of Γ is determined by the series of future and past observations, or, in other words, two distinct doubly infinite series of past and future observations correspond to distinct phase points).

(b) Information relative to observations in the infinitely distant future is trivial information, that is either almost certain or almost impossible for the equilibrium probability μ, i.e.

$$\bigwedge_{-\infty}^{+\infty} T^i = \Omega$$

A dynamical system having these properties is called a Kolmogorov system. We see that it is so unstable as to be unpredictable in the distant future whatever information we may have about its evolution over past times, however long. The evolution of a system of hard spheres is of this type [16]. The partition $\mathscr{P}$ corresponds to an observable whose distant future becomes more or less independent of the past. Clearly, not every partition $\mathscr{P}$ possesses the above two properties and this is why the functional (11) is defined as a function of such a partition.

One can now formulate the second principle of thermodynamics as a selection rule for the physical states of the dynamical system under consideration.

We shall regard as physically admissible the set of probability distributions v having finite information $\Omega_\mu(v, \mathscr{P})$ with respect to μ. Let $\mathscr{D}(\mathscr{P})$ denote this class of distributions.

One can then prove the following results [4, 5]:

(a) The class $\mathscr{D}(\mathscr{P})$ is not invariant under inversion of time.
(b) If the system is initially in one of the states v of $\mathscr{D}(\mathscr{P})$, then it will tend, over the course of its future dynamic evolution, towards the equilibrium μ.
(c) If v_t is the distribution (carried by T) at the instant t then

$$S_\mu(v_t, \mathscr{P}) - -\Omega_\mu(v_t, \mathscr{P})$$

is a negative quantity that increases monotonically with time. It tends to zero as $t \to +\infty$. This value is only attained by the equilibrium state μ.

The first property (a) allows one to express the intrinsic irreversibility of the dynamical system. Such irreversibility cannot manifest itself at the level of the description of states of the system by phase points. However, we now know that trajectories are unstable in sufficiently complicated dynamical systems as this follows from the theory of Lyapounov exponents (on this subject, see the work of Arnold and Avez [1] and of Lichtenberg and Lieberman [12]) which restricts the validity of the dynamic description. Thus, as soon as we pass to the statistical description, property (a) implies the possibility of the fracture of temporal symmetry. In fact, the operation of time inversion (generated by the operation of velocity inversion in classical mechanics) is characterized by the two properties: I is a transformation of Γ into Γ such that $I(I\omega) = \omega$ and $ISI^{-1} = S^{-1}$. One can show [5] that $I\,\mathscr{D}(\mathscr{P})$ is a class having the same properties as $\mathscr{D}(\mathscr{P})$ for $t \leqslant 0$. It follows from this that there exist states that tend to equilibrium as $t \to \infty$ but not as $t \to -\infty$, and that, under inversion of time, behave in the opposite way.

How then is the second principle of thermodynamics to be expressed? Above all, this principle expresses an impossibility. At the level of statistical states it expresses itself through the impossibility of getting states that do not tend towards equilibrium in the future (of the form $I\mathscr{D}(\mathscr{P})$). By way of illustration, one can consider the impossibility of coming across distributions describing a set of particles moving in a parallel formation after multiple collisions with fixed obstacles, while it is possible to have an initially parallel beam of particles that then go off in various directions following multiple collisions.

Property (b) expresses the principle of the approach to equilibrium. Property (c) then expresses the principle of the increase of entropy or the decrease of 'negentropy', here regarded as the excess of information contained in the state with respect to μ.

18.3 Concluding remarks

The presentation given here is sufficiently abstract and general to provide results in the setting of the kinetic theory of particles interacting through collisions. However, the real difficulties of this theory [3] fully justify the search for new ways in the theory of irreversibility and thermodynamic entropy. What we have looked at here is essentially the formulation of the principle of entropy. It is well known that the problem with the kinetic theory is the derivation of the Boltzmann-type irreversible equations associated with dense gases. Construction of the equations of this type is possible via a theory by which a Markov semi-group is associated with every unstable system. We refer the reader to [8] and [13].

Lastly, the exclusion of certain states from $I\,\mathscr{D}(\mathscr{D})$ corresponds to the exclusion of states with infinite information. In fact, since the $\mathscr{D}(\mathscr{P})$ is not invariant under the inversion of time there then exist states $\hat{v} = Iv$, $v \in \mathscr{D}(\mathscr{P})$, such that $\Omega_\mu(\bar{v}, \mathscr{P}) = +\infty$. It follows from this that the inversion of time corresponds, for certain states, to an infinite flux of information, and it is the impossibility of such a flux that is expressed as the selection rule for physically admissible states.

Note This work comprises part of a research programme led by the researchers of the Brussels group.

Acknowledgement The author would like to express his gratitude for the fruitful collaboration that he has had with the members of this group.

References

1. Arnold V. I. and Avez A. (1968) *Ergodic Problems of Classical Mechanics*, New York.
2. Brillouin L. (1963) *Science and Information Theory*, Academic Press, New York.
3. Cohen E. G. D. (1983) Kinetic theory of non-equilibrium fluids, *Physica*, **118A**, 17.
4. Courbage M. (1983) Intrinsic irreversibility of Kolmogorov dynamical systems, *Physica*, **122A**, 459–482.

5. Courbage M. and Prigogine I. (1983) Intrinsic randomness and intrinsic irreversibility in classical dynamical systems, *Proc. Nat. Acad. Sci., USA*, **80**, 2412.
6. Dobrushin R. L. (1963) *General Formulation of Shanon Basic Theorems of the Theory of Information*, English Translation in American Mathematical Society Translations, Series 2, Vol. 33.
7. Gelfand I. M. and Yaglom A. Y. (1959) *Calculation of the Amount of Information about a Random Function Contained in Another Such Function*, English translation in American Mathematical Society Translations, Providence R.I., Series 2, Vol. 12.
8. Goldstein S., Misra B. and Courbage M. (1981) *J. Statist. Phys.*, **25**, 111.
9. Khinchin A. I. (1957) *Mathematical Foundations of Information Theory*, New York.
10. Klein M. J. (1973) The development of Boltzmann statistical ideas, in E. G. D. Cohen and Thirring (Eds.), *The Boltzmann Equation, Theory and Applications*, Springer-Verlag, p. 53.
11. Kullback S. (1959) *Information Theory and Statistics*, New York.
12. Lichtenberg J. and Lieberman M. A. (1983) *Regular and Stochastic Motion*, Springer-Verlag.
13. Misra B., Prigogine I. and Courbage M. (1979) *Physica*, **98A**, 1–26.
14. Penrose O. (1979) Foundations of statistical mechanics, *Rep. Prog. Phys.*, **42**, 1937–2006.
15. Perez A. (1957) Notions généralisées d'incertitude, d'entropie, et d'information, in *Transactions of the First Prague Conference on Information Theory etc.*, Czech. Acad. Sci., Prague, pp. 183–208.
16. Sinai Ya. G. (1970) Dynamical systems with elastic reflexions, English Transl. in *Russ. Math. Surv.*, **25**, 137–189.
17. Yaglom A. I. and Yaglom M. I. (1969) *Probabilité et Information*, Dunod, Paris.

Chapter 19 Dimension and entropy of regular curves

M. Mendès-France
Department of Mathematics, University of Bordeaux I

19.1 By way of introduction

I first met B. Mandelbrot in 1967 seated in a cafe in the Latin quarter. He explained to me that, as a specialist in 'paradoxical cases', he had discovered why the empirical formulae of geographers led to poor predictions. These formulae took no account of the dimension of the object under study; this had to do with the coast of Brittany whose dimension exceeds 1. The introduction of Hausdorff-type dimensions into the natural sciences has fascinated me from that moment, but I did not see how I could contribute to this new science. When, several years later, Mandelbrot's first book [7] appeared, I was sufficiently mature to read and appreciate it. It is certainly one of the books that have made the strongest impression on me. I would here like to illustrate at least two of the ideas that grew out of it and which struck me the most forcibly.

Although a mathematician, Mandelbrot refuses to give a precise definition of fractal. Moreover, he gives several non-equivalent definitions of the notion of dimension. One can well understand why. The object of study, whether a coastline, an outline, a galaxy, a language, price fluctuation, etc., is given, is inflexible. Mathematics, on the other hand, is supple. One can, indeed one should, fit the mathematical definition to the nature of the object studied. What one cannot do, although too often one would like to, is to reduce it to a mathematical definition, which is necessarily too narrow to take account of its precise nature. It is for this reason that each problem should entail the choice of an appropriate notion of dimension. There is no single dimension; there are as many as one wants. In the study of plane curves, which I intend to discuss here, I will thus choose a definition of the dimension suited to the problem.

The second idea of Mandelbrot that I would like to illustrate here is that dimension is a measure of complexity. I will restrict myself to plane curves which are locally rectifiable in the plane. I therefore exclude Weierstrass curves, brownian curves and Peano curves, which will certainly be discussed in other chapters. For the curves that concern me, I will define a dimension d, $1 \leqslant d \leqslant 2$ (d cannot exceed 2, because everything takes place in the plane), and an entropy

H, $0 \leqslant H \leqslant 1$. It is well known that the entropy measures the complexity. I will show that, for a large class of curves.

$$d \geqslant \frac{1}{1-H}$$

which clearly expresses the fact that the dimension increases with the entropy.

In the interests of simplification, I have taken it upon myself to modify certain definitions that I have given in previous articles. I think I have made it clear enough why I should be allowed to do this. I am quite certain that any reader, far from holding it against me, will gladly condone it.

19.2 Dimensions of plane curves [2, 9]

Consider a plane curve Γ of infinite length and such that any bounded set in the plane contains no more than a finite portion of Γ. It follows that the curve must tend to the point at infinity in the plane, and such a curve will be called regular. Examples are a straight line, a spiral $\rho = f(\theta)$ where $f(\theta)$ tends to infinity with θ, $y = \sin x$, etc.

Before defining the dimension $d(\Gamma)$ we recall two simple facts. Consider an infinite straight line Δ and a disc D_R with centre at the origin and radius R (Figure 19.1). Let $\Delta(R) = \Delta \cap D_R$ be the part of Δ interior to D_R and let $|\Delta(R)|$ be its length. When R tends to infinity

$$|\Delta(R)| \sim 2R \tag{1}$$

Further, let Σ be one of the half-planes bounded by Δ. Put $\Sigma(R) = \Sigma \cap D_R$ and denote its area by $|\Sigma(R)|$. (The symbol $|\cdot|$ always stands for the 'natural' measure. If x is a finite set, $|x|$ is its cardinality; if x is a curve, $|x|$ is its length; and, lastly, if x is a surface, $|x|$ is its area.) When R increases indefinitely

$$|\Sigma(R)| \sim \tfrac{1}{2}\pi R^2 \tag{2}$$

The exponents of R in the formulae (1) and (2) respectively reflect the topological dimension of Δ, on the one hand, and of Σ, on the other, whence the idea of

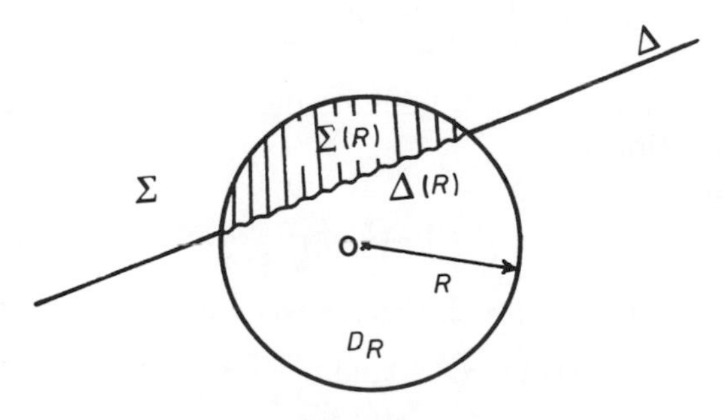

Figure 19.1

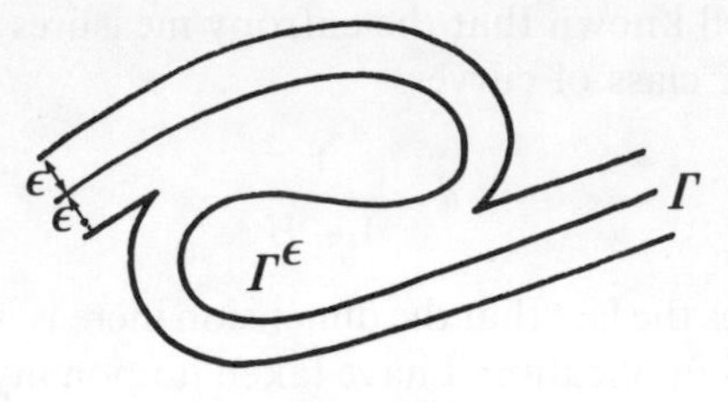

Figure 19.2

defining the dimension of a curve Γ by the exponent d in the formula

$$|\Gamma \cap D_R| = |\Gamma(R)| \sim cR^d$$

More precisely, one would define the dimension of Γ by means of the (provisional) formula

$$d = \lim_{R\to\infty} \frac{\log|\Gamma(R)|}{\log R} \tag{3}$$

This formula is open to criticism on at least two grounds. To start with, the limit can fail to exist. One can then rather take a limit superior or inferior. More serious is the fact that, if Γ is very convoluted, formula (3) can very well lead to a value of $d > 2$. Now it stands to reason that every plane object should have dimension $d \leqslant 2$. It is therefore necessary to modify the definition.

Let $\varepsilon > 0$. Let Γ^ε be the 'ε-enlargement' of Γ (Figure 19.2):

$$\Gamma^\varepsilon = \{p \in \mathbf{R}^2 : \mathrm{dist}(p, \Gamma) < \varepsilon\}$$

Γ^ε is thus a surface. Let, then, $\Gamma^\varepsilon(R)$ be the part of Γ^ε interior to the disc D_R. The upper (respectively lower) dimension is given by

$$\bar{d}(\Gamma) = \lim_{\varepsilon\to 0} \lim_{R\to\infty} \sup \frac{\log|\Gamma^\varepsilon(R)|}{\log R}$$

$$\underline{d}(\Gamma) = \lim_{\varepsilon\to 0} \lim_{R\to\infty} \inf \frac{\log|\Gamma^\varepsilon(R)|}{\log R}$$

If the two quantities are equal, we denote their common value by $d(\Gamma)$, which is called the dimension Γ.

It is clear that

$$1 \leqslant \underline{d}(\Gamma) \leqslant \bar{d}(\Gamma) \leqslant 2$$

and for all α, β with $1 \leqslant \alpha \leqslant \beta \leqslant 2$, there exists a curve Γ such that

$$\underline{d}(\Gamma) = \alpha \qquad \text{and} \qquad \bar{d}(\Gamma) = \beta$$

By way of example, one can convince oneself that a straight line is of dimension

1, that the spiral $\rho = \theta^\alpha$ where $\alpha > 0$ has dimension

$$\min\left\{2, 1 + \frac{1}{\alpha}\right\}$$

that the spiral $\rho = \exp\theta$ has dimension 1 and that, lastly, the spiral $\rho = \log(1 + \theta)$ has dimension 2.

19.3 Soluble curves

Let Γ_s be the initial section of Γ of length s (we are thus assuming that an origin for the abcissae has been chosen on Γ). Let Γ_s^ε be the ε-enlargement of Γ_s. Γ is said to be soluble if there exists $\varepsilon > 0$ such that

$$\liminf_{s \to \infty} \frac{|\Gamma_s^\varepsilon|}{s} > 0$$

A straight line Γ is clearly soluble because

$$|\Gamma_s^\varepsilon| \sim 2\varepsilon s$$

when s tends to infinity. Similarly for the spiral $\rho = \theta$. On the other hand, a curve that would tend to collapse on itself, such as $\rho = \log(\theta + 1)$ or $y = \sin x^2$, is not soluble.

If Γ is soluble, then for every $\varepsilon > 0$ and for R tending to infinity,

$$0 < A(\varepsilon) < \frac{|\Gamma^\varepsilon(R)|}{|\Gamma(R)|} < B(\varepsilon)$$

and so the dimensions $\bar{d}(\Gamma)$ and $\underline{d}(\Gamma)$ can be expressed more simply as

$$\bar{d}(\Gamma) = \limsup_{R \to \infty} \frac{\log|\Gamma(R)|}{\log R}$$

$$\underline{d}(\Gamma) = \liminf_{R \to \infty} \frac{\log|\Gamma(R)|}{\log R}$$

(Unsurprisingly, we have recovered definition (3)). The dimension of a soluble curve is thus easy to calculate.

If, on the other hand, the curve is not soluble,

$$\bar{d}(\Gamma) \geqslant \limsup_{R \to \infty} \frac{\log|\Gamma(R)|}{\log R}$$

$$\underline{d}(\Gamma) \leqslant \limsup_{R \to \infty} \frac{\log|\Gamma(R)|}{\log R}$$

An insoluble curve is thus more complicated to study. We shall see, moreover,

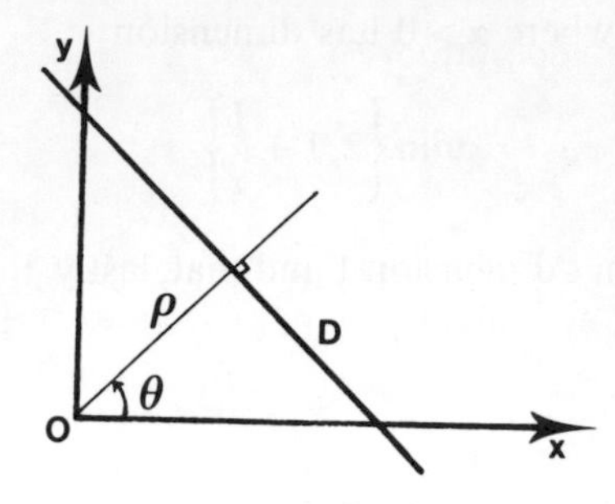

Figure 19.3

that every curve of large entropy is necessarily insoluble. What the entropy is will be explained in the following sections.

19.4 Entropy of finite curves

Let $\Gamma = \Gamma_s$ be a curve of length s. Let $\Omega(\Gamma)$ be the set of straight lines D that cut Γ. A straight line D is generally determined by its distance $\rho \geqslant 0$ from the origin, together with its direction coefficients $\cos\theta$, $\sin\theta$ (Figure 19.3).

In the (ρ, θ) plane, $\rho \geqslant 0$, $0 \leqslant \theta < 2\pi$, we are interested in the set $\Omega(\Gamma)$ endowed with the uniform probability measure

$$\frac{\mathrm{d}\theta\,\mathrm{d}\rho}{\operatorname{meas}\Omega(\Gamma)}$$

The set of straight lines intersecting Γ is thus endowed with a probability law. Let p_n denote the probability that a straight line D cuts Γ in exactly n points $(1 \leqslant n)$. H. Steinhaus has established that the mean number of points of intersection of Γ with a random straight line is

$$\sum_{n=1}^{\infty} n\,p_n = \frac{2|\Gamma|}{|\delta k|}$$

Here, $|\Gamma|$ denotes the length of Γ $(=s)$ and $|\delta k|$ is the length of the frontier of the convex hull k of Γ (Figure 19.4).

One will observe that, for every finite curve, $|\delta k| \leqslant 2|\Gamma|$, with equality occurring only for straight line segments.

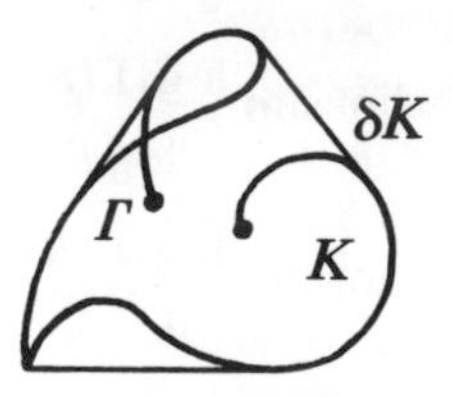

Figure 19.4

The entropy of Γ is by definition

$$S(\Gamma)=\sum_{n=1}^{\infty} p_n \log \frac{1}{p_n}$$

with the convention that $0\cdot\infty=0$. The entropy is either positive or zero. $S(\Gamma)$ is only zero if Γ is a straight line segment ($p_1=1$, $p_n=0$ for $n\geqslant 2$) or a closed convex curve ($p_2=1$, $p_n=0$ for $n\neq 2$). To a complicated curve there corresponds a high entropy.

Theorem 1 For every finite curve

$$S(\Gamma)\leqslant \log\frac{2|\Gamma|}{|\delta k|}+\frac{\beta}{e^{\beta}-1}$$

where

$$\beta=\log\frac{2|\Gamma|}{2|\Gamma|-|\delta k|}\geqslant 0 \qquad \text{and} \qquad 0\leqslant\frac{\beta}{e^{\beta}-1}\leqslant 1$$

Proof Consider the entropy function

$$S(x_1,x_2,\ldots)=\sum_{n=1}^{\infty} x_n \log\frac{1}{x_n}$$

A classical elementary calculation (Lagrange multipliers) enables one to find the maximum of S, subject to the variables being related by the two conditions

$$\sum_{n=1}^{\infty} x_n=1, \qquad \sum_{n=1}^{\infty} n x_n=\frac{2|\Gamma|}{|\delta k|}$$

The maximum obtained is

$$S_{\max}=\log\frac{2|\Gamma|}{|\delta k|}+\frac{\beta}{e^{\beta}-1}$$

(In two previous articles, I have explained how to interpret the coefficient β: it is the inverse of a temperature. One thus defines the temperature of a curve [4], [8]. We do not develop this theme here.)

19.5 Entropy of an infinite curve

Let Γ be an infinite curve and let Γ_s be the initial segment of length s. From the preceding section, we know what the entropy of Γ_s is. We define the relative entropy as

$$\frac{S(\Gamma_s)}{\log|\Gamma_s|}=\frac{S(\Gamma_s)}{\log s}$$

The upper (respectively lower) relative entropy of the curve Γ is then

$$\bar{H}(\Gamma) = \limsup_{s\to\infty} \frac{S(\Gamma_s)}{\log s}$$

$$\underline{H}(\Gamma) = \liminf_{s\to\infty} \frac{S(\Gamma_s)}{\log s}$$

When these two quantities are equal, we denote their common value by $H(\Gamma)$. The preceding theorem shows that

$$0 \leqslant \underline{H}(\Gamma) \leqslant \bar{H}(\Gamma) \leqslant 1$$

and, further, it is easy to see that given α and β such that $0 \leqslant \alpha \leqslant \beta \leqslant 1$, there exists an infinite curve Γ with

$$\underline{H}(\Gamma) = \alpha \qquad \text{and} \qquad \bar{H}(\Gamma) = \beta$$

By definition, an infinite curve Γ is said to be determinist if $H(\Gamma) = 0$. It is chaotic if $H(\Gamma) = 1$. Let us give some examples. An infinite straight line, an exponential spiral $\rho = e^\theta$ and the sine curve $y = \sin x$ are all determinist. The spiral $\rho = \theta^\alpha$ ($\alpha > 0$) has entropy $1/(1+\alpha)$. The spiral $\rho = \log(\theta + 1)$ and every bounded curve of infinite length are all chaotic. We observe that the spiral $\rho = \log(\theta + 1)$ strongly resembles the form of certain vortices associated with turbulence. Such a spiral thus has every right to be regarded as chaotic! On the other hand, the spiral $\rho = e^\theta$ of zero entropy is reminiscent of biological growths (a snail's shell) and should be associated with order and determinism.

Let $\theta_1, \theta_2, \theta_3, \ldots$ be an infinite sequence of independent random variables uniformly distributed on $(0, 1)$. The entropy H of the curve consisting of the polygonal line whose nth vertex has coordinates

$$x_n = \cos 2\pi\theta_n, \qquad y_n = \sin 2\pi\theta_n$$

is almost surely $\frac{1}{2}$. This type of polygonal line has been especially studied by Lehmer [5], Dekking and myself [2], Deshouillers [3], Callot and Diener [1] and Loxton [6].

We will now demonstrate the connection between entropy and dimension.

Theorem 2 If Γ is soluble,

$$\bar{H}(\Gamma) \leqslant 1 - \frac{1}{\bar{d}(\Gamma)}, \qquad \underline{H}(\Gamma) \leqslant 1 - \frac{1}{\underline{d}(\Gamma)}$$

If, further, Γ is of dimension one, then Γ is determinist.

Proof Suppose that Γ is soluble. Then

$$\bar{d}(\Gamma) = \limsup_{R\to\infty} \frac{\log|\Gamma(R)|}{\log R} = \limsup_{s\to\infty} \frac{\log s}{\log|\delta k_s|}$$

Moreover, by Theorem 1,

$$\bar{H}(\Gamma) \leqslant \limsup_{s\to\infty} \frac{\log 2s/|\delta k_s|}{\log s} = 1 - \frac{1}{\limsup \log s/(\log|\delta k_s|)}$$

$$\leqslant 1 - \frac{1}{\bar{d}(\Gamma)}$$

Similarly,

$$\underline{H}(\Gamma) \leqslant 1 - \frac{1}{\underline{d}(\Gamma)}$$

If, further, Γ is of dimension 1, then $\bar{d}(\Gamma) = 1$, and so $\bar{H}(\Gamma) = 0$. QED

Corollary If $\bar{H}[\Gamma) > \frac{1}{2}$, then Γ is insoluble.

Proof Suppose that $\bar{H}(\Gamma) > \frac{1}{2}$ and that Γ is soluble. By Theorem 2,

$$\tfrac{1}{2} < \bar{H}(\Gamma) \leqslant 1 - \frac{1}{\bar{d}(\Gamma)}$$

and so $\bar{d}(\Gamma) > 2$, which is absurd. Hence Γ is insoluble. QED

This last result seems very suggestive to me. $\bar{H}(\Gamma) > \frac{1}{2}$ actually implies a large entropy and therefore a certain complexity. Insolubility, moreover, implies obscurity and confusion. The corollary thus implies that complexity entails confusion.

This rather ordinary comment endows, *a posteriori*, the chosen definitions with a certain reality. I will therefore venture to claim to have proved (or maybe tested?) my definitions.

References

1. Callot J. L. and Diener M. (1984) *Variations en Spirales*, Oran.
2. Dekking F. M. and Mendès-France M. (1981) Uniform distribution modulo one: a geometrical viewpoint, *J. Reine Angew. Math.*, **329**, 143–153
3. Deshouillers J. M. (1984) *Geometric Aspect of Weyl Sums, Elementary and Analytic Theory of Numbers*, Banach Centre Publications, pp. 75–82.
4. Dupain Y., Kamae J. and Mendès-France M. (1986) Can one measure the temperature of a curve? *Archive for Rational Mechanics and Analysis*. **94**, 115–163; (1987) *Corrigenda*, **98**, 395.
5. Lehmer D. H. (1976) Incomplete Gauss sums, *Mathematika*, **23**, 125–135.
6. Loxton J. (1984) The graphs of exponential sums, *Mathematika*.
7. Mandelbrot B. (1975) 1st ed. (1984) 2nd ed., *Les Objects Fractals*, Flammarion. See also *The Fractal Geometry of Nature*, Freeman, 1982.

8. Mendès-France M. (1984) Folding paper and thermodynamics, Physics report 103, pp. 161–172.
9. Mendès-France M. and Tenenbaum G. (1981) Dimension des courbes planes, papiers pliés et suites de Rudin-Shapiro, *Bull. Soc. Math. Fr.*, **109**, 207–213.
10. Santalo L. (1976) *Integral Geometry and Geometric Probability.* Encyclopedia of Mathematics, Addison Wesley.
11. Steinhaus H. (1954) Length, shape and area, *Colloquium Mathematicum*, **3**, 1–13.

Chapter 20 Local dimension, momentum and trajectories

Guy Cherbit
Research Group in Membrane Biophysics, University of Paris VII

From Platon to the general relativity of A. Einstein [3], there have been numerous attempts to geometrize the physical universe.

In his articles on entropic invariance in electrochemistry, A. Le Méhauté [5, 6] even associates a fractal dimension with the flux of entropy across the surface of an electrode. In fact, this flux is expressed as a function of a non-integral exponent, closely related to the complexity of the surface of the porous electrode. This is not because this exponent is equal to the dimension of the surface of the electrode. Nevertheless, this formulation is interesting in that it geometrizes a scalar quantity, the entropy.

Recall that for B. Mandelbrot, a fractal is strictly geometrical [7]. However, we have already discussed this restriction.

With great elegance and daring, M. Mendès-France goes in the opposite direction and 'physicalizes' geometry, even going so far as to speak of the entropy and even the temperature of a curve! We emphasize that the approach is quantitative and allows one to measure the degree of complexity of a curve (or, which amounts to the same thing, the indistinguishability of its component parts).

The transfer to physics is immediate. One simply regards the curve as being the trajectory of a particle. This transfer only looks easy and immediately raises fundamental questions concerning the notion of non-integral dimension:

(a) Is it a *relative* notion? Consider for example, the movement of a paramecium in a stagnant pond (Figure 20.1). The entropy (in the sense of Mendès-France) of its trajectory is very high and its dimension is close to 2.

If one considers the same disturbance not just in still water, but in the current of a stream, the trajectory (relative to the bed) is stretched. Its entropy diminishes and its dimension approaches 1. This example clearly shows that the dimension of a trajectory depends on the state of motion of the moving object. Moreover, adopting the terminology of Mendès-France, the positional uncertainty Δx (i.e. the indistinguishability of parts of the curve and therefore the dimension) depends on the momentum of the particle

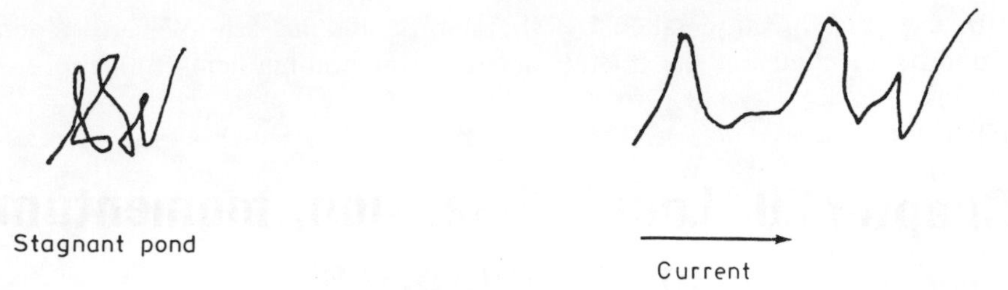

Figure 20.1

through the uncertainty relation:

$$\Delta p \, \Delta x = h$$

The dimension is thus dependent on the frame of reference of the observer.

(b) Is this a *subjective* notion? One can ask the question of whether or not the dimension *also* depends on the observer. Indeed, parts of the curve which are indistinguishable to the naked eye may no longer be so when viewed through the microscope. The dimension, then, depends on the yardstick of the observer, i.e. on the choice, made by the observer, of 'reading pace'.

It is then the observer who fractalizes the object and the dimension is no more than the measure of the sharpness of the experimenter's observation.

This is the problem with fractal measures made on photographs. It is essential to distinguish clearly the different degrees of enlargement to which the plate can be subjected, as well as, in microphotography, the magnification of the microscope.

Mathematicians, of course, do not consider this problem since they pass to the limit in making the 'reading pace' tend to zero.

For practical purposes, this passage to the limit is impossible and inevitably leads to the definition of a domain of validity for the fractal dimension.

This cut-off point is often artificial and leads to a modelling that is sometimes far removed from the reality. When an electron (elementary particle level) changes energy level in DNA cell tissue (molecular level) at the time of reproduction of the cell, this latter is going to generate an abnormal cell (cell level) which in its turn will generate other abnormal cells which will subsequently form a cancerous tissue (tissue level). This can then swarm through the entire organism from the abnormal cells. This invasion can then lead to death (individual level).

This oversimplified scenario nevertheless shows that no meaningful level of scale can be specified *a priori* in the example given.

If the object exhibits an invariance of scale under internal similarity, this problem does not arise, but I know of no concrete example of such an object. Tired of this frequently raised objection, Mandelbrot has recently replaced self-similarity by self-affinity [8]. This is more flexible but rare all the same.

In fact, the direction taken by Mandelbrot and his school has decided upon subjectivity of the fractal dimension in non-mathematical applications.

(c) Is this an *intrinsic* notion? This question is, in fact, crucial. As the work on fractals has developed, the tool has been promoted (making it an end in itself) at the expense of the physical notion of non-integral dimension. This aspect will be developed in the next chapter on 'spatiotemporal fractalization'.

For the time being, let us return to our trajectories: the *dimension of the trajectory depends on the momentum.*

We have just seen that this comes about via the uncertainty relation. We know that every moving object having momentum p is 'piloted' by an associated wave:

$$\lambda = \frac{h}{p}$$

The moving object can thus only be located to within λ. This is the lower limit of the 'indistinguishability' of the parts of the trajectory.

The dimension thus depends on λ and therefore on the momentum p of the particle.

To characterize the trajectory T in an intrinsic way, one could, among other methods, use Kolmogorov's notion of ε-entropy:

$$H_\varepsilon(T) = \log_2 N_\varepsilon(T)$$

(where $N_\varepsilon(T)$ denotes the minimal number of sets of diameter ε needed to cover the trajectory T), taking, for example, $\varepsilon = \inf(\lambda)$. Unfortunately, p is not bounded and $\inf(\lambda) = 0$.

One comes back to the problem, mentioned above, of passage to the zero limit. In other words, if one hopes to define an intrinsic dimension in applications where passage to the zero limit is not possible, it is necessary to admit the existence of a *natural* lower limit for the measurement gauge [2]. Certain consequences of this choice will be discussed in the next chapter.

Let us return to the impulse–dimension relationship. In numerous articles, it is shown in various ways that the dimension of the trajectory of a brownian motion is $D = 2$ [9]. The same methods are used to show that a quantum trajectory also has dimension equal to 2.

Going back to 1966, Feynmann [4] had already pointed out that a quantum trajectory is continuous but not differentiable.

The problem of the dimension of such a trajectory is treated in detail by Abbott and Wise [1]. If one measures the position of the particle to within ε at instants Δt apart, its wave function may be written as

$$\psi_\varepsilon(x) = \frac{\varepsilon^{3/2}}{\hbar^3} \int_{\mathbf{R}^3} \frac{1}{(2\pi)^{3/2}} f\left(\frac{|p|\varepsilon}{\hbar}\right) e^{ip(x/\hbar)}\, d^3p$$

where f is chosen so that the position of the particle is specified to within ε.

If $\bar{l}$ denotes the mean distance covered in time Δt, we have, as $\psi\psi^*$ measures the probability density of finding the particle at x during the period Δt:

$$\bar{l} = \int_{\mathbf{R}^3} |x| \, |\psi_\varepsilon(x, \Delta t)|^2 \, \mathrm{d}^3x$$

Let $\bar{L} = N\bar{l}\varepsilon^{D-1}$ be the fractal length associated with the trajectory after N observations. Averaging a transformation on f and putting $x = k\varepsilon$, one obtains

$$\bar{L} = |\bar{p}| \frac{t}{m} \int_{\mathbf{R}^3} |F(\alpha, \beta)|^2 \left| \frac{\bar{p}}{|\bar{p}|} + \frac{\hbar k}{2|\bar{p}|\varepsilon\beta} \right| \varepsilon^{D-1} \, \mathrm{d}^3\alpha$$

where t measures the total duration of the observations of the particle of mass m, and where $\bar{p}$ denotes the mean momentum.

Let $\bar{\lambda} = h/|\bar{p}|$ denote the mean wavelength associated with the particle. Several cases can then occur:

(a) If the observation gauge ε is very large compared with $\bar{\lambda}$ (we are then in the setting of classical mechanics), then

$$\frac{\hbar}{|\bar{p}|\varepsilon} \text{ is very small}$$

and

$$\bar{L} \simeq \frac{\bar{p}}{m} t \int_{\mathbf{R}^3} |F(\alpha, \beta)|^2 \varepsilon^{D-1} \, \mathrm{d}^3\alpha$$

For $\bar{L}$ to be independent of ε, it is necessary that $D = 1$. One then recovers the classical result.

(b) If ε is small compared with $\bar{\lambda}$ (the case of quantum mechanics)

$$\frac{\hbar}{|\bar{p}|\varepsilon} \text{ is very large compared with } \frac{\bar{p}}{|\bar{p}|}$$

and

$$\bar{L} \simeq \frac{\bar{p}}{m} t \int_{\mathbf{R}^3} |F(\alpha, \beta)|^2 \left| \frac{\hbar k}{2|\bar{p}|\varepsilon\beta} \right| \varepsilon^{D-1} \, \mathrm{d}^3\alpha$$

For $\bar{L}$ to be independent of ε, it is then necessary that $D = 2$. We thus recover the quantum result.

(c) For intermediate situations, i.e. when ε is arbitrary compared with $\bar{\lambda}$, one can only say that with ε given, D depends on p and that $1 < D < 2$.

One of course recovers the impulse–dimension dependance, which thus appears to be essential to the study of trajectories.

The Heisenberg inequalities render the concept of trajectory inadequate. Its complexity increases with resolution and goes from being perfectly smooth ($D = 1$) to $D = 2$ in the quantum case, the 'anfractuosities' increasing as D passes from 1 to 2. The quantum study of the microscopic complexity of a trajectory reveals the non-differentiability of this trajectory. It then becomes impossible to keep the definition of velocity:

$$\mathbf{v} = \frac{\mathbf{ds}}{\mathrm{d}t}$$

One therefore has to generalize the notion of velocity and associate with it a local definition of the dimension. This is the direction in which we are currently working (Cherbit, Jonot and Billot), and we now present the essence of the method followed.

20.1 Local dimension—generalized velocity

20.1.1 α-Differentiability

Let $X(t)$ be the position vector of a particle at the instant t. X is said to be α-differentiable at t if there exists a function u that is positively homogeneous of order α such that

$$\lim_{s\to 0} \frac{|X(t+s) - X(t) - u(s)|}{|s|^{\alpha}} = 0$$

note that u is positively homogeneous of order α when

$$\forall t \in \mathbf{R} \forall x \in \mathbf{R}^3 u(tx) = |t|^{\alpha} u(x)$$

The two quantities $X(t+s) - X(t)$ and $u(s)$ (Figure 20.2) thus have contact of order α at $s = 0$. Note that if u exists, it is unique. We put

$$u = D_t^{(\alpha)} X$$

20.1.2 Local dimension

The following properties are easily established:

(a) If X is continuous at t, then X is zero-differentiable.
(b) If X is d-differentiable at t, then X is α-differentiable for every $\alpha < d$ and $D_t^{(\alpha)} X = 0$.

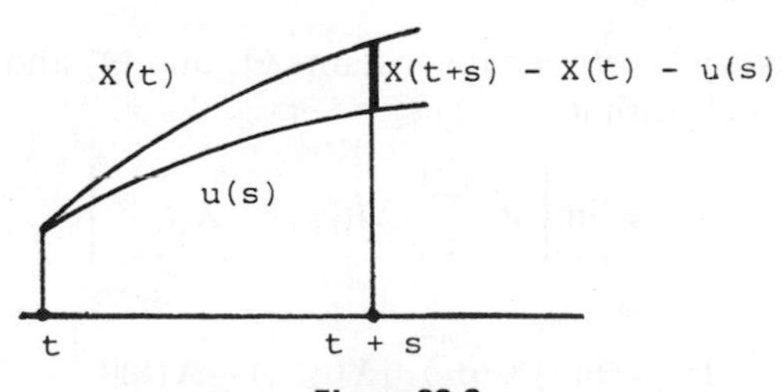

Figure 20.2

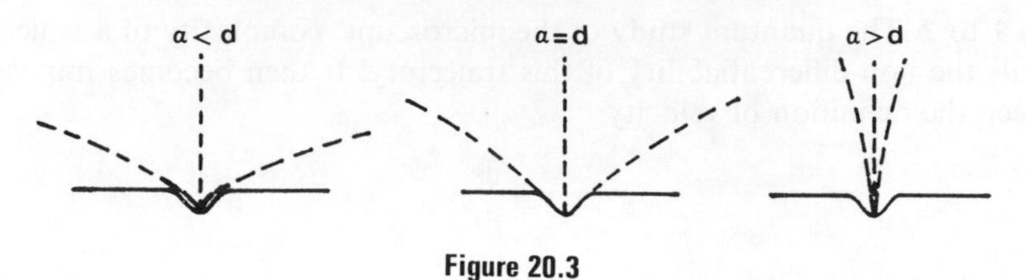

Figure 20.3

(c) We can therefore compile the following table:

α	0	d ∞
$D_t^{(\alpha)}X$	0	Undefined

This table defines the *local dimension* d of X at the instant t.

This means that one 'tests' the anfractuosity at the instant t with the help of points $|s|^\alpha$; d is the value beginning with which the 'point' no longer dovetails into the anfractuosity at t (Figure 20.3).

Let us go back to the definition. The initial idea is the following: a straight line (the tangent, if it exists) does not hold enough information to 'read off' what is going on at a point, unless the curve C is very smooth there.

20.1.3 α-Velocity

By definition, the α-velocity on the trajectory is

$$X^{(\alpha)}(t) = \lim_{s\to 0} \frac{X(t+s) - X(t)}{s^\alpha} \qquad \text{where } \alpha \in (0, 1]$$

the time being endowed with the metric defined by the valuation $|s|^\alpha$.

Of course, this α-velocity at the instant t is equal to $u(1)$. When $\alpha = 1$, one recovers the classical notion of velocity.

20.1.4 The notion of *D*-length

Consider the trajectory Γ between the instants Θ_1 and Θ_2 and dissect $[\Theta_1, \Theta_2]$ into intervals $[t_i, t_{i+1}]$, with $|t_{i+1} - t_i| \leqslant \varepsilon$. Let

$$l^\alpha_{\min} = \lim_{\varepsilon\to 0}\left[\inf \sum_{i=0}^{n-1} |X(t_{i+1}) - X(t_i)|^\alpha\right]$$

and

$$l^\alpha_{\max} = \lim_{\varepsilon\to 0}\left[\sup \sum_{i=0}^{n-1} |X(t_{i+1}) - X(t_i)|^\alpha\right]$$

We shall say that Γ is α-measurable when

$$l^{\alpha}_{\min} = l^{\alpha}_{\max}$$

Now suppose that Γ is uniformly continuous between the instants Θ_1 and Θ_2 and let

a = the maximal value of α for which $l^{\alpha}_{\min}$ remains finite

b = the maximal value of α for which $l^{\alpha}_{\max}$ remains finite

When $a=b$, we call $D=a=b$ the dimension of the trajectory between Θ_1 and Θ_2. This new dimension corresponds to a *filling* of the space by the trajectory.

Similarly, $l^D = l^D_{\min} = l^D_{\max}$ defines the D-length of Γ traversed between the instants Θ_1 and Θ_2.

Theorem 1 If $X(t)$ is $1/\alpha$-differentiable and α-measurable betwee Θ_1 and Θ_2, then

$$l^{\alpha} = \int_{\Theta_1}^{\Theta_2} |X^{(1/\alpha)}(t)|^{\alpha}\, \mathrm{d}t$$

Theorem 2 If X is $1/\alpha$-differentiable on an interval I, then

$$\dim_H X(I) \leqslant \dim_T X(I) \leqslant \alpha$$

where $\dim_H$ and $\dim_T$ stand for the Hausdorff and packing dimensions respectively of $X(I)$.

20.2 Summary

To sum up, we have defined a local dimension via the notion of α-differentiability. From it, we have derived the expression for the velocity at each point.

On an arc of the trajectory, the filling dimension corresponds to the maximal local complexity and allows one to calculate a 'filling length'. This leads us quite naturally to define the dimension of a physical space $\mathcal{U}$ as being the greatest dimension attained by every trajectory of this space. This dimension is intrinsic.

In this space, one can then rewrite the uncertainty relations, but these now involve the dimension.

However that may be, this chapter illustrates well the thinking that lay at the origin of the seminars and of this work.

Out of mathematics of the past (Hausdorff, etc.), new ideas and methods (Bouligand, Lévy, Kahane, Peyrière, etc.) have grown. Parallel to this, Mandelbrot has authoritatively shown that non-integral dimensions are very widespread in numerous areas and has suggested a new way of looking at nature. It is then quite legitimate to ask whether an 'object is fractalized by the universe to which it belongs or rather by the analysis of the observer?'.

Whatever interest this question may hold, this leads us to make new mathematical statements and thus to develop new tools and sometimes even new concepts.

Herein lies the essence of this work.

References

1. Abbott L. F. and Wise W. B. (1981) Dimension of a quantum-mechanical path, *Am. J. Phys.*, **49** (1), 37.
2. Cherbit G. Séminaire Hausdorff, June 1985, Spatio-temporal fractalisation.
3. Einstein A. (1953) *Geometry and Experience*, Gauthier-Villars.
4. Feymann R. P. and Hibbs A. R. (1965) *Quantum Mechanics and Path Integrals*, McGraw-Hill, New York.
5. Le Méhauté A. (1982) *C.R. Acad. Sci., Paris*, **294**, 836, Series II.
6. Le Méhauté A. (1986) Séminaire Hausdorff, March 1986, Chaos and entropy (G. Cherbit).
7. Mandelbrot B. Séminaire Hausdorff, June 1985, Spatio-temporal fractalisation (G. Cherbit).
8. Mandelbrot B. Séminaire Hausdorff, January 1986, Catalysis (P. Pfeifer).
9. Nelson E. (1966) *Phys. Rev.*, **150**, 1079.

Chapter 21 Space–time dimensionality

Guy Cherbit
Research Group in Membrane Biophysics, University of Paris VII

Spatio-temporal fractalizations are comparatively rare. G. N. Ord [3] has published a very complete study, which includes relativistic quantum mechanics.

The body of theory is very elegant but at the cost of an artifice of calculation that can seem rather awkward. In fact, Ord uses Peano curves in a five-dimensional hyperspace: three dimensions for the space, one for the real time and one for the imaginary time. The paths along this last axis are interpreted as the apparition of an anti-particle. This raises certain contradictions in connection with the real arrow of time but, in fact, this is of no great importance for the bulk of the work which remains very attractive.

Of course, the Peano–Moore curves are not here physical trajectories but a simple tool (which, besides, the author very soon disposes of by making the resolution step tend to zero). The generalization, by analogy, to a general trajectory is a little hasty, but the step is fundamental: *the particle moves in a fractal space-time.*

Now we have seen, in the previous chapter, that the dimension of a trajectory is a relative notion (related to the momentum of the particle). This implies that to each moving body there is attached a fractal space-time, whose trajectory would be a geodesic.

Moreover, we have already pointed out the difficulty of resorting to mathematical definitions (because of the zero limit) for applications to other disciplines. For the results to retain an, albeit relative, intrinsic aspect, one can, for example, allow the resolution steps which are used to study different trajectories of the same space-time to be multiples of the same unit quantity [1]. Let us make this point precise.

In field theory, the success of gauge theories on a discrete lattice [5] is well known. A gauge theory generalizes a field in assuring the invariance of physical results under local transformations of the potential. For example, Maxwell's equations are invariant under the transformation

$$V(x) \to V(x) + \frac{\partial}{\partial t}\lambda(x)$$

$$\mathbf{A}(x) \to \mathbf{A}(x) - \mathbf{grad}\ \lambda(x)$$

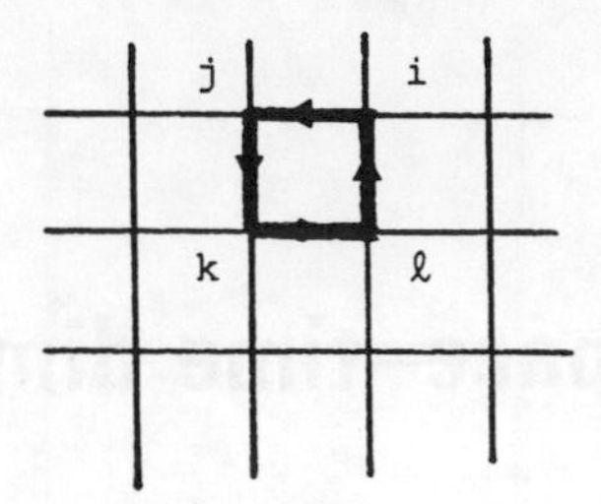

Figure 21.1 With each link ij one associates a complex number R_{ij} of modulus 1 depending on the ij component of the four-vector (V, **A**)

The gauge theories on a discrete lattice use a space-time whose coordinates can only take discrete values corresponding to the sites of a lattice (Figure 21.1).

The interest of this lies in its making it possible to use all the methods of statistical mechanics.

Of course, the model should be gauge-invariant and reduce to the continuous theory when the mesh of the lattice tends to zero.

Suppose here that:

(a) To each particle is associated a discretized space-time.
(b) There exists a lower bound (δ, τ) to the spatio-temporal mesh of the lattice.
(c) In this discretized space-time, the trajectories are fractal because the geodesics of the space are (Figure 21.2).

Consider then a particle in uniform translation with respect to a reference state called 'at rest'. From the point of view of this reference state, the dimension of the space associated with the particle depends on its momentum (Figure 21.3) and is given by

$$D = n_0 - \varphi(p)$$

where n_0 denotes the dimension of the space linked to the particle as seen from the particle itself.

The function φ should satisfy the boundary conditions

$$\varphi(0) = 0 \qquad \text{and} \qquad \varphi(\infty) = 1$$

and, moreover, satisfy

$$\varphi(p + p') = 1 \qquad \text{if } \varphi(p) = 1$$

One easily shows that $\varphi(p)$ is then a solution of the equation

$$\varphi(p + p') = \frac{\varphi(p) + \varphi(p')}{1 + \varphi(p)\varphi(p')}$$

(note the resemblance to relative addition of velocities) and that

$$\varphi(p) = \text{th}\, \alpha p \qquad \text{where } \alpha = \varphi'(0)$$

It follows from the quantification (δ, τ) of the space and the time that the

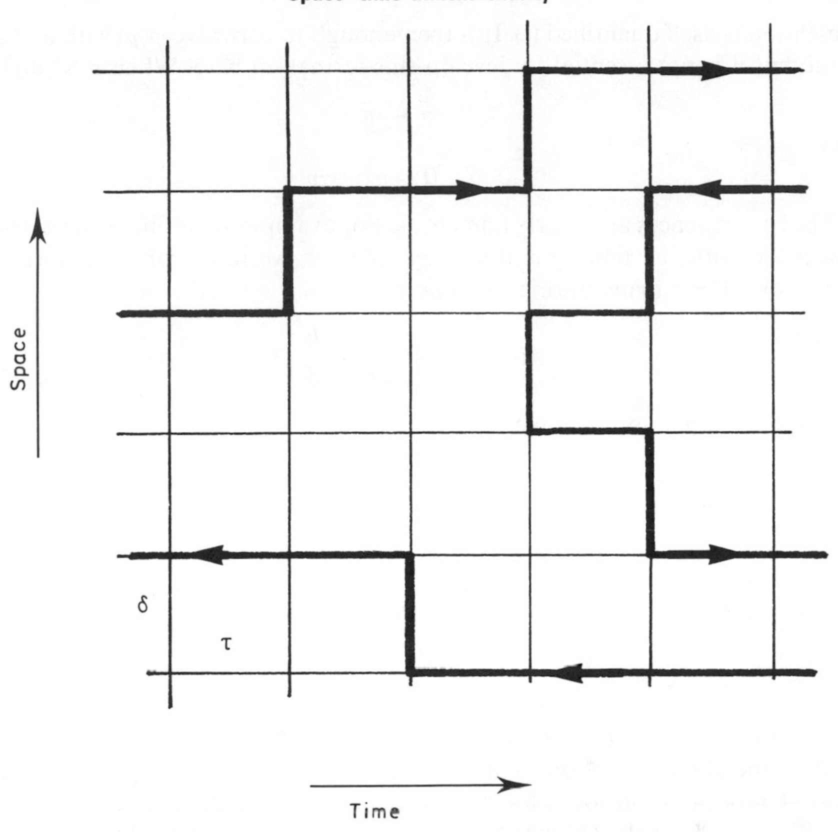

Figure 21.2 Unresolved trajectory in a discretized space-time (disintegration of a meson)

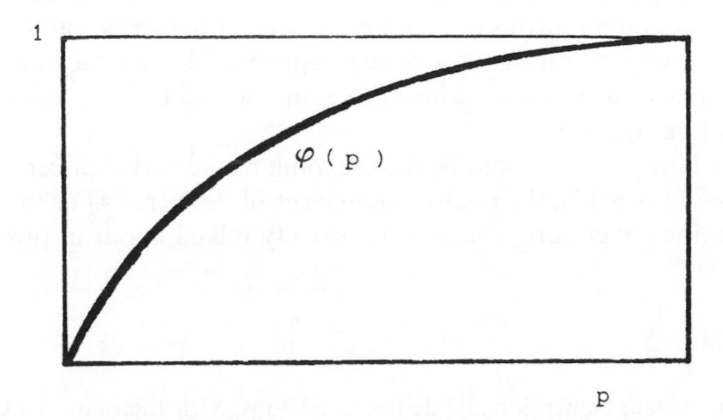

Figure 21.3 Variation of dimension with momentum

dimension is itself quantified (ε). It is then enough to convolve $\varphi(p)$ with a Dirac comb but it is not essential to develop this formalism here. When p is small,

$$\text{th}\,\alpha p \sim \alpha p$$

and

$$D \sim n_0 - \alpha p$$

The consequences are clearly numerous. For example, denoting the frequency associated with the time quantum τ by $\nu_0 = 1/\tau$, we have, for an elementary transition $\mathrm{d}D = \varepsilon$ made during the time τ:

$$\varepsilon = \alpha \frac{h\nu_0}{c} = \alpha \frac{h}{c\tau} = \alpha \frac{h}{\delta}$$

From this, we get, for example,

$$c = \frac{\alpha h}{\varepsilon \tau}$$

The fact that one cannot exceed the velocity of light c therefore follows from the quanta ε *and* τ (or ε and δ).

Observe that we also recover the *uncertainty relation*

$$\delta \frac{\varepsilon}{\alpha} = h$$

since ε/α is the momentum necessary for the transition ε.

This uncertainty relation is interpreted as the impossibility of having a spatio-temporal localization less than the mesh (δ, τ) of the lattice.

Without too much difficulty, one can reconstruct an extended relativistic quantum mechanics on the basis of the above work, but I do not intend to do this here.

I wanted to show that behind the notion of fractal object there is, in embryo, the possibility of introducing the new concept of non-integral physical dimension and that this leads almost inexorably to a universal mesh-limit to the distinguishability of the parts of a physical object. In such a space, Kolmogorov's δ-entropy takes on an intrinsic sense and a path in the space can be described in a binomial way by means of a binary sequence. (Under the condition that one works in a system of units where $c = 1$, one would then have $\delta = \tau$ and the mesh would be square.)

In such a lattice, what would be the meaning of the folded paper of Dekking and Mendès-France [2], the random sequences of Peyrière [4] or the sequences of perturbations of cultures that I have already talked about in this work?

References

1. Cherbit G. (1986) Dimensionalité de l'espace-temps, XIth International Congress of Cybernetics, Namur.

2. Dekking M., Mendès-France M and Vander Poorten (1982) Folds! in *The Mathematical Intelligencer*, Vol. 4, Springer-Verlag.
3. Ord G. N. (1983) Fractal space-time: a geometric analogue of relativistic quantum mechanics, *J. Phys. A: Math. Gen.*, **16** (9) 1869–1884.
4. Peyrière J. (1985) Séminaire Hausdorff, Représentation de fractal par des measures aléatoires.
5. Wilson K. (1976) *Phys. Reports, C,* **23** (3), 331–347.

Hausdorff seminars

Organized by the Biophysics Research Group, University of Paris VII

—*Fractals and Materials*, February 1985
A. Le Méhauté, M. Keddam, D. Sweich, L. Fruchter

—*Stochastic processes and Hausdorff dimension*, March 1985
M. Weber, M. P. Assouad, J. Peyrière

—*Fractal modelling in biology*, April 1985
J. P. Rigaut, M. Marcovich

—*Models of irregular curves*, May 1985
S. Dubuc, M. Mendès-France, M. Dekking, G. Deslauriers

—*The concept of non-integral dimension in physics*, June 1985
G. Cherbit, B. Mandelbrot, J. Chanu, A. Le Méhauté

—*Representation of fractals by random measures*, December 1985
J.-P. Kahane, M. Hugonin, J. Peyrière, M. Rosso

—*Catalysis and fractal surfaces*, January 1986
P. Pfeifer, M. Keddam, H. Takenouti, H. Van Damme

—*Ising model, automata and fractals*, February 1986
M. Mendès-France, J. P. Allouche, B. Derrida

—*Chaos and entropy*, March 1986
G. Cherbit, M. Courbage, A. le Méhauté

Secretary: P. Brandouy

Index

Index compiled by G. C. Jones